Systems of Equations

Systems of Equations

Word Problems
and
Step-by-Step Solutions

Arben Alimi

Systems of Equations
Word Problems and Step-by-Step Solutions

Copyright © 2016 Arben Alimi

Library of Congress Control Number: 2016902626

ISBN: 1523465344

ISBN-13: 978-1523465347

CreateSpace Independent Publishing Platform, North Charleston, SC

Acknowledgments

It is with ineffable gratitude that I acknowledge the immense support and help of my family in writing this book. I owe my deepest gratitude to my caring and loving wife Sevim for managing household and other activities while I was trying to write, calculate and complete this book. I am indebted to my seven year old son Artrim for showing understanding that daddy is busy and does not have enough time to play with him. Finally, the completion of this book would not have been accomplished without the tremendous academic support of my daughter Trina. Her insightful suggestions, meticulous editing and proofreading made this book possible. I share the credit of my work with Trina.

Contents

Introduction to Systems of Equations 1

1. Introduction to Substitution Method 4

2. Introduction to Addition Method.. 8

3. Fixing one coefficient 10

4. Fixing both Coefficients... 13

5. Brushing up Substitution Method... 16

6. Brushing up Addition method 19

7. Fractions example 21

8. Square roots example... 24

9. Reinforcing Addition Method 27

10. Reinforcing Substitution Method 29

11. Introducing the Graph method.. 32

12. Reinforcing Graph method 36

13. The sum and difference of two numbers.................................... 40

14. The sum and product of two numbers 43

15. The relationship between two numbers...................................... 48

16. Chocolates and Candies 51

17. Age difference 55

18. Restaurant TIPS 58

19. Bikes 62

20. Apple devices.. 66

21. Mortgage and Property Taxes 71

22. Planning for a Vacation... 74

23. How many questions on a test 77

24. The two-digit number 80

25. The three-digit number 84

26. The speed of Hudson River 92

27. The speed of a Mountain River 95

28. Investing in stocks 98

29. College tuition 102

30. Diluting full fat milk 106

31. Movie popularity 110

32. Internet speed 114

33. Deciding which Car to buy 118

34. Should you get a Master's Degree? 123

35. Watch your Calorie intake 129

36. Employee earnings 134

37. Employee and Manager Raises 139

38. Get the Money ready 144

39. Student loans 147

40. Study time for SAT 151

41. Basketball shots 155

42. Basketball shooting competition 158

43. Basketball free-throws 161

44. Number of Coins in pocket 165

45. Number of Coins in wallet 168

46. Legs of a Triangle 172

47. Shipment Quantity 176

48. Business Break Even Point 182

49. How to choose a phone carrier 185

50. Pizza with toppings 189

Preface

You should consider this book as a Personal Tutor who is sitting next to you and is trying hard to make it easy and fun while you are honing your skills on Systems of Equations. This "Personal Tutor" will not only help you understand and be able to solve Systems of Equations, but will influence your thinking as well. As a good Tutor would do, this book does not simply give you the solutions, but explains in detail the thinking behind every step as the examples are being solved. As a result, you will gain confidence in solving Systems of Equations and use this knowledge to solve real world problems in everyday life. This book-tutor is an affordable alternative to the Real Tutors. This book will be with you all the time and will go over an example as many times as you want, any time you want.

This book is a fruit of father-daughter Math collaboration. As Trina was studying Systems of Equations, and was trying to solve various Word Problem examples, very often she needed a confirmation that she got the Equations right and they truly represent the problem as it is described in the question. Some of the Math problems were trivial, but some were tricky and required some serious thinking.

As a result of our Math conversations and the fruits that came out of it, we decided to share our methodology with you. We came up with new examples, wrote down what we were thinking as we solved the problems and put them in this book, so that you and every other student that comes across can benefit the same way Trina benefited from these detailed explanations. Every example in this book is explained in such a way so that it made sense to Trina, the 8-th grade student. Her understanding is the testimony that every solution of the 50 examples is easy to follow, detailed, and fully explained.

Being a computer programmer myself, I always think in terms of algorithms, where every step is clearly defined, and not a single step skipped, so that non-intelligent machines, such as computers, can "solve problems" and appear to be intelligent. Similarly, every example in this book is solved step-by-step without a step skipped, so that the student can understand every detail while a Math Word Problem is "translated" into a Mathematical model and solved using the Systems of Equations.

The first 12 Examples will explain various methods for solving Systems of Equations. Once armed with this knowledge and skills, the rest of the examples become easy to follow. Learning the techniques first is very important, because when the Word Problems are elaborated, the Student is focused on the thinking process instead of techniques of solving the System of Equations. Correct thinking and the way you approach the problem are the most important ingredient while constructing the Mathematical Models, which in our case are the Equations of the System of Equations. Once the System is constructed, finding the solutions is easy.

You can go over the examples in any order because they are fully explained and do not depend on the knowledge gained from previous examples. But we still think it is better to go over them in order because we feel that the first ones are easier and can help you build a sound thinking foundation and be better prepared for the later examples, which we feel are more complex.

Introduction to Systems of Equations

Before we try to explain what **Systems of Equations** are, it will be very helpful to understand what an Equation is, what a System is, and what do they mean or represent in real life.

We all know what an equation looks like. However, many of us cannot clearly articulate what an Equation is and what it represents in real life. We all know that an Equation has x-s and y-s and if we are asked to write one, we all would write something like $2x + 5 = y$ or $x^2 + 3x - 1 = 0$. Furthermore, most of us know how to algebraically manipulate and solve them, but if we are asked where we use them, most of us would say, unfortunately, in School or a Math class to solve Math problems and pass the class. While all Equations represent some Real World phenomena, we fail to recognize them as such, and we don't make use of them in our everyday lives.

Equations get their name from the Equal Sign $(=)$ they contain in them. Recall that if we have something with x-s and y-s and there is no equal sign, then it is called an Algebraic Expression, or simply an Expression. The equal sign of an Equation tells us that the "thing" on the left, which is an Expression, is equal in value to the "thing" or the Expression on the right. No matter how simple, or how complex or how different the expressions on the left and on the right of the equal sign might look like, they must be equal in value if we claim that the Equation has a solution. If values for which the left and the right side become equal cannot be found, then we say that the Equation has no solution.

Now that we understand the Algebraic notation of Equations, let's say few words about what they mean or what they represent. In the simplest terms we can say that an Equation is a **statement** about a specific relationship of some natural phenomena. The individual natural phenomena that we are analyzing usually have names, for example an English word, or we can simply assign a letter such x or y to each of them. If two or more natural phenomena have some sort of relationship, then that relationship usually can be described in English sentences, or we can use letters such x or y to describe that relationship. If we were to describe the relationship of two or more natural phenomena while using letters such x or y, then we would use Math operators such as $+, -, \times, \div$ etc. to connect them and thus make an Expression. If we were to describe some specific property of this Expression or to quantify it, then we would use the equal sign $(=)$.

Let's clarify all of this through a Real World example. Let's say we are talking about Nina and her brother Artie. Many things can be said about them, but in this moment, let' say that we are analyzing their Ages. So, Age is the natural phenomena under our consideration. We can use English sentences and write that Nina is "this" many years old and Artie is "that" many years old. To name their Ages, we can use the letter x to mean Nina's age and y to mean Artie's age. But what if we want to describe some relationship between their ages? Well, then we will use some Math operator and form an Expression. If we are going to describe how old they are together we would add their Ages and write an Expression like $x + y$. If we are going to describe the fact that Nina is older than Artie and would like to know their Age difference, then we would write an Expression like $x - y$. If we would like to describe the specific fact that Nina and Artie together are **10** years old, then we would write an Equation: $x + y = \mathbf{10}$. If we would like to describe the specific fact that Nina is **7** years older than Artie, then we would write an Equation like $x - y = \mathbf{7}$. If we would like to describe the specific fact that Nina is two times older than Artie, then we would write $x = \mathbf{2}y$.

Let us now explain the meaning of the word System. Whenever we consider things, natural phenomena, parts, or elements etc., and we create a relationship or interdependency among them and thus view them as whole or as connected things, know that we have formed a System of those things, elements, phenomena or parts. We have created a Group or a Set of tightly connected things, which are somehow depended on each other. Otherwise, if those things are not somehow connected, or they don't depend on each other, have nothing in common, no relationship with each other, then they wouldn't form a System.

Now it is easy for us and we are ready to describe what a System of Equations is and what it represents.

A System of Equations is a group of two or more Equations that form a System. Since they form a System, it means that those Equations have some relationship with each other, they are somehow connected, they form a coherent whole, and they are interdependent. Their relationship, connectedness, and interdependency can be seen from the fact that they use the same set of variables (individual natural phenomena). The fact that those Equations form a coherent whole means that you deal with them all at once, they are inseparable and that's why they form a System.

Since we are clear now what System of Equations are, let's wrap this discussion with an explanation of what they represent in real world. Whenever we describe two or more specific relationships of natural phenomena and we would like to quantify those phenomena, we make a System of Equations and solve it.

If this last sentence is not fully clear, rest assured that by the end of this book it will be crystal clear. This is the very purpose of this book. Almost all examples of this book describe a Real World situation, describe some relationship of some phenomena and provide step-by-step solutions until those phenomena are quantified.

1. Introduction to Substitution Method

Solve for **x** and **y**:

$$\begin{cases} x + y = 3 \\ x - y = 1 \end{cases}$$

Solution

Since this is the first example on "Systems of Equations", then before we find the solutions of this very simple example of Systems of Equations, let's make sure that we understand what the word "Systems" means.

You already know that equations can exist on their own. Each of the two equations above can simply exist on its own and it will not depend on the other. Take for example the second equation $(x - y = 1)$. You can find unlimited pairs (x, y) so that $x - y = 1$. In plain English this means that you can find **x** and **y** so that when you subtract **y** from **x** the result is **1**. Such pairs would be $(2, 1)$, $(3, 2)$, $(4, 3)$, $(5, 4)$, …. . However, once we add another condition and we say that "when you add **x** and **y** their sum is **3**", then not all the pairs we just listed before will satisfy this second condition.

So, the first equation tells us one condition that **x** and **y** must satisfy, and the second equation tells us another condition that **x** and **y** must satisfy. These two conditions have a strong relationship with each other. They are tied together and cannot be separated. Both equations make a cohesive whole. In one word, they form a "System" and that's why we call them "System of Equations".

In the above "System of Equations" you are being asked to find such an **x** and **y** so that they satisfy the following two conditions:

1. When you add **x** and **y** the result is **3**, AND
2. When you subtract **y** from **x** the result is **1**.

Finding such values for x and y so that the above-mentioned two conditions are met means solving the System of Equations.

How do we solve the above System of Equations?

Well, there are few ways to solve it. Here we will look at **"Substitution Method"**. Since the two equations are related, we can simply isolate one variable in one equation and **substitute** it in the other equation. That is why this method of solving the Systems of Equations is called **"Substitution Method"**.
So, we have:

$$\begin{cases} x + y = 3 \\ x - y = 1 \end{cases}$$

Let's pick the second equation and isolate **x**:

$$\begin{cases} x + y = 3 \\ x = 1 + y \end{cases}$$

Let's substitute **x** in the first equation with the value of **x** in the second equation which is $\mathbf{1 + y}$:

$$x + y = 3$$

$$(1 + y) + y = 3$$

Let's solve for **y** :

$$1 + y + y = 3$$

$$1 + 2y = 3$$

$$2y = 3 - 1$$

$$2y = 2$$

$$y = \frac{2}{2}$$

$$y = 1$$

Now that we found the value of **y**, we can substitute it in any equation and find the value of **x**. Let's substitute it in the first equation:

$$x + y = 3$$
$$x + 1 = 3$$
$$x = 3 - 1$$
$$x = 2$$

So, the solutions for our System of equations are:

$$\mathbf{x = 2} \quad \text{and} \quad \mathbf{y = 1}$$

Check your work

Since it is very easy to make mistakes when dealing with numbers, it is very important to be able to prove that our result is correct. We found in this example and claim that **x = 2** and **y = 1** are the solutions of our System of equations, but are we correct?

How do we prove that the values of our variables are correct?

The way to prove that our calculated values are correct is to plug them in the System of Equations and see that all equations are still in balance, meaning that the left side of equations equals the right side of equation. If they are not in balance, then it means that somewhere during calculation we went wrong and we need to correct it.

Let us check now that our result **x = 2** and **y = 1** is a correct one for our System of Equations. We will substitute these values in the System and see whether our equations are still in balance.

Our System of Equations is:

$$\begin{cases} x + y = 3 \\ x - y = 1 \end{cases}$$

After we plug in the values for **x** and **y**:

$$\begin{cases} 2 + 1 = 3 \\ 2 - 1 = 1 \end{cases}$$

$$\begin{cases} 3 = 3 \\ 1 = 1 \end{cases}$$

Since our equations are in balance, this means that our result was correct.

2. Introduction to Addition Method

Solve for **x** and **y** using Addition Method:

$$\begin{cases} x + y = 3 \\ x - y = 1 \end{cases}$$

Solution

As we learned in the first example, one method for solving the Systems of Equations was the **"Substitution Method"**. Another method is the **"Addition Method"**.

It is called **"Addition method"** because we add the two equations. The way to add them is to add the **left** side of first equation to the **left** side of the second equation, and the **right** side of the first equation to the **right** side of the second equation.

The **Addition method** is also called **Elimination method** because when we add the two equations we eliminate one of the two variables and solve for the remaining one. In other words, the reason why we add the two equations is to eliminate one of the variables and find the value of the other.

So, we have:

$$\begin{cases} x + y = 3 \\ x - y = 1 \end{cases}$$

Let's add the sides of equations:

$$(x + y) + (x - y) = 3 + 1$$

Notice how **y** will disappear after we add the "like terms"!

$$x + y + x - y = 3 + 1$$

$$x + x = 3 + 1$$

$$2x = 4$$

$$x = \frac{4}{2}$$

$$x = 2$$

Now that we found the value of **x**, we can substitute it in any equation and find the value of **y**. Let's substitute it in the first equation:

$$x + y = 3$$
$$2 + y = 3$$
$$y = 3 - 2$$
$$y = 1$$

So, the solutions for our System of equations are:

$$x = 2 \quad \text{and} \quad y = 1$$

Check your work

The way we check our work and prove that our results are in fact the correct solution for our System of Equation is by plugging in the values of the variables in the equations and see that they are still in balance.

Our System of Equations is:

$$\begin{cases} x + y = 3 \\ x - y = 1 \end{cases}$$

After we substitute the values:

$$\begin{cases} 2 + 1 = 3 \\ 2 - 1 = 1 \end{cases}$$
$$\begin{cases} 3 = 3 \\ 1 = 1 \end{cases}$$

Since our equations are still in balance, this means that our result was correct.

3. Fixing one coefficient

Solve for **x** and **y** using the Addition Method:

$$\begin{cases} 3x + y = 7 \\ 2x + y = 5 \end{cases}$$

Solution

As we learned in the previous example, when we use the "Addition method" we simply add the sides of the equations. We added the left side of first equation to the left side of the second equation, and the right side of the first equation to the right side of the second equation.

We also said that the **Addition method** is also called **Elimination method** because when we add the two equations we eliminate one of the two variables and solve for the one which is left.

However, it will not be always the case that one of the variables will be eliminated by simply adding the equations. If we add the above equations we will get $\mathbf{5x + 2y = 12}$, and this will not help us in any way to find the values of **x** and **y**.

In order to eliminate one of the variables, we need to **modify one or both** equations in a way so that we get the same coefficient in front of one of the variables in the two equations but with opposite signs $(+, -)$.

How do we modify the equations?

You need to remember this:

1. If you multiply the left and the right side of an equation by the same number, the equation will still remain in balance. It will just look differently.
2. Use your imagination and creativity to find one or two numbers to make the coefficients in front of one of the variables the same value but with opposite signs.

If you take a closer look, you can see that the coefficient of **y** is **1** in both equations, which makes things easier a little bit. If you multiply, let's say, the second equation by $(-\mathbf{1})$, the equation will still be in balance but the **y** in the second equation will have the opposite sign of the **y** in the first equation. This will make possible to eliminate **y** when you add the equations. So, we have:

$$\begin{cases} 3x + y = 7 \\ 2x + y = 5 \end{cases}$$

Multiply the second equation by $(-\mathbf{1})$:

$$\begin{cases} 3x + y = 7 \\ 2x + y = 5 \qquad / \times (-1) \end{cases}$$

$$\begin{cases} 3x + y = 7 \\ -2x - y = -5 \end{cases}$$

Let's add the left sides of equations, and right sides as well:

$$(3x + y) + (-2x - y) = 7 + (-5)$$

Notice how **y** disappears (cancels out) after we remove the parenthesis and we add the terms!

$$3x + y - 2x - y = 7 - 5$$

$$3x - 2x = 2$$

$$x = 2$$

Now that we found the value of **x**, we can substitute it in any equation and find the value of **y**. Let's substitute it in the second equation:

$$2x + y = 5$$

$$2 \times 2 + y = 5$$

$$4 + y = 5$$

$$y = 5 - 4$$

$$y = 1$$

So, the solutions for our System of Equations are:

$$x = 2 \quad \text{and} \quad y = 1$$

Check your work

The way we check our work and prove that our results are in fact the correct solution for our System of Equations is by plugging in the values of the variables in the equations and see that they are still in balance.

Our System of Equations is:

$$\begin{cases} 3x + y = 7 \\ 2x + y = 5 \end{cases}$$

After value substitution:

$$\begin{cases} 3 \times 2 + 1 = 7 \\ 2 \times 2 + 1 = 5 \end{cases}$$

$$\begin{cases} 7 = 7 \\ 5 = 5 \end{cases}$$

Since our equations are still in balance, this means that our result was correct.

4. Fixing both Coefficients

Solve for **x** and **y** using the Addition Method:

$$\begin{cases} 4x + 3y = 11 \\ 3x + 2y = 8 \end{cases}$$

Solution

As we learned in the previous examples, when we use the "Addition method" we simply add the two equations. The way to add them is to add the left side of first equation to the left side of the second equation, and the right side of the first equation to the right side of the second equation.

The **Addition method** is also called **Elimination method** because when we add the two equations we eliminate one of the two variables and solve for the one that remained.

However, not always will one of the variables be eliminated just by adding the equations. If we add the above equations we will get $7x + 5y = 19$, and this will not help us in any way to find the values of **x** and **y**.

In order to eliminate one of the variables, we need modify one or both equations in a way so that we get the same coefficient in front of one of the variables but with opposite signs $(+, -)$.

How to modify the equations then?

You need to remember this:

1. If you multiply the left and the right side of an equation by the same number, the equation will still remain in balance. It will just look differently, but they will stay identical.
2. Use your imagination and creativity to find one or two numbers to make the coefficients in front of one of the variables the same value but with opposite signs.

In the previous example it was enough to modify only one equation and we got the opposite signs or coefficients $(+, -)$ in front of one of the variables. However, in this

System of equations that is not possible. We have to modify both equations and bring the coefficients of one of the variables to the same value but with opposite signs.

If you take a closer look, you can see that if you decide to eliminate the variable **y**, one way would be to multiply the first equation by **2** and you will get **6y**, and multiply the second equation by (-3) and you will get $(-6y)$. This will make it possible to eliminate **y** when you add the equations.

So, we have:

$$\begin{cases} 4x + 3y = 11 \\ 3x + 2y = 8 \end{cases}$$

Multiply the first equation by **2**, and second equation by (-3):

$$\begin{cases} 4x + 3y = 11 & / \times 2 \\ 3x + 2y = 8 & / \times (-3) \end{cases}$$

$$\begin{cases} 8x + 6y = 22 \\ -9x - 6y = -24 \end{cases}$$

Let's add the left sides of equations, and right sides as well:

$$(8x + 6y) + (-9x - 6y) = 22 + (-24)$$

Notice how **y** disappears (cancels out) after we remove the parenthesis and add the terms:

$$8x + 6y - 9x - 6y = 22 - 24$$
$$8x - 9x = -2$$
$$-x = -2$$

Multiply both sides by (-1) to remove the negative sign:

$$-\mathbf{x} = -2 \quad / \times (-1)$$
$$\mathbf{x} = 2$$

Now that we found the value of **x**, we can substitute it in any equation and find the value of **y**. Let's substitute it in the second equation:

$$3x + 2y = 8$$

$$3 \times 2 + 2y = 8$$

$$6 + 2y = 8$$

$$2y = 8 - 6$$

$$2y = 2$$

$$y = \frac{2}{2}$$

$$y = 1$$

So, the solutions for our System of equations are:

$$x = 2 \quad \text{and} \quad y = 1$$

Check your work

The way we check our work and prove that our results are in fact the correct solution for our System of Equations is by plugging in the values of the variables in the equations and see that they are still in balance.

Our System of Equations is:

$$\begin{cases} 4x + 3y = 11 \\ 3x + 2y = 8 \end{cases}$$

After value substitution:

$$\begin{cases} 4 \times 2 + 3 \times 1 = 11 \\ 3 \times 2 + 2 \times 1 = 8 \end{cases}$$

$$\begin{cases} 11 = 11 \\ 8 = 8 \end{cases}$$

Since our equations are still in balance, it means that our result was correct.

5. Brushing up Substitution Method

Solve for **x** and **y** by using the **Substitution** Method:

$$\begin{cases} 4x + 3y = 11 \\ 3x + 2y = 8 \end{cases}$$

Solution

Note that we already solved this system of equations in the previous example using the "Addition method". This time we will solve it using the "Substitution Method". Know that whichever method you choose to solve the system of equations, the result will be the same. You should choose whichever method is easier for you and whichever method makes solving the system of equations simpler.

Recall that it is called "**Substitution Method**" because we pick one equation and solve (isolate) for one of the variables, and then we **substitute** it in the other equation with its value (usually an expression). Let's demonstrate this in action.

Let's pick the first equation and solve for **x**:

$$4x + 3y = 11$$

$$4x = 11 - 3y$$

$$x = \frac{11}{4} - \frac{3}{4}y$$

Let's substitute the value of this **x** in the second equation. So, the second equation is:

$$3x + 2y = 8$$

After substitution the equation looks:

$$3 \times \left(\frac{11}{4} - \frac{3}{4}y\right) + 2y = 8$$

Let's solve for **y**:

$$\frac{33}{4} - \frac{9}{4}y + 2y = 8$$

Let's multiply the equation by **4** so that we remove the denominator of the fractions and make our life easier when solving this equation:

$$\frac{33}{4} - \frac{9}{4}y + 2y = 8 \qquad / \times 4$$

$$4 \times \left(\frac{33}{4} - \frac{9}{4}y + 2y\right) = 4 \times 8$$

$$4 \times \frac{33}{4} - 4 \times \frac{9}{4}y + 4 \times 2y = 4 \times 8$$

$$33 - 9y + 8y = 32$$

$$-9y + 8y = 32 - 33$$

$$-y = -1$$

After we multiply both sides by (-1) to remove the negative sign:

$$y = 1$$

Now that we found the value of **y**, we can substitute it in any equation and find the value of **x**. Let's substitute it in the second equation because the numbers are smaller, which makes it easier to work with:

$$3x + 2y = 8$$

$$3x + 2 \times 1 = 8$$

$$3x + 2 = 8$$

$$3x = 8 - 2$$

$$3x = 6$$

$$x = \frac{6}{3}$$

$$x = 2$$

So, the solutions for our System of equations are:

$$x = 2 \quad \text{and} \quad y = 1$$

Check your work

The way we check our work and prove that our results are in fact the correct solution for our System of Equations is by plugging in the values of the variables in the equations and see that they are still in balance.

Our System of Equations is:

$$\begin{cases} 4x + 3y = 11 \\ 3x + 2y = 8 \end{cases}$$

After value substitution:

$$\begin{cases} 4 \times 2 + 3 \times 1 = 11 \\ 3 \times 2 + 2 \times 1 = 8 \end{cases}$$

$$\begin{cases} 11 = 11 \\ 8 = 8 \end{cases}$$

Since our equations are still in balance, it means that our result was correct.

6. Brushing up Addition method

Solve for **x** and **y** :

$$\begin{cases} 2x + 5y = -9 \\ 3x + 3y = 0 \end{cases}$$

Solution

Let's use the Addition method to find the solutions. So, we have:

$$\begin{cases} 2x + 5y = -9 \\ 3x + 3y = 0 \end{cases}$$

Multiply the first equation by **3** and second equation by (-2) to eliminate **x**:

$$\begin{cases} 2x + 5y = -9 & / \times 3 \\ 3x + 3y = 0 & / \times (-2) \end{cases}$$

$$\begin{cases} 6x + 15y = -27 \\ -6x - 6y = 0 \end{cases}$$

Let's add the left side of the first equation to left side of the second equation, and the right sides as well:

$$(6x + 15y) + (-6x - 6y) = -27 + 0$$

Let's remove the parenthesis and add the like terms:

$$6x + 15y - 6x - 6y = -27$$

$$9y = -27$$

$$y = -\frac{27}{9}$$

$$y = -3$$

Now that we found the value of **y**, we can substitute it in any equation and find the value of **x**. Let's substitute it in the second equation:

$$3x + 3y = 0$$

$$3x + 3 \times (-3) = 0$$

$$3x - 9 = 0$$

$$3x = 0 + 9$$

$$3x = 9$$

$$x = \frac{9}{3}$$

$$x = 3$$

So, the solutions for our System of Equations are:

$$x = 3 \text{ and } y = -3$$

Check your work

Let's plug in the values of the variables in the equations and see whether they are still in balance.

Our System of Equations is:

$$\begin{cases} 2x + 5y = -9 \\ 3x + 3y = 0 \end{cases}$$

After value substitution:

$$\begin{cases} 2 \times 3 + 5 \times (-3) = -9 \\ 3 \times 3 + 3 \times (-3) = 0 \end{cases}$$

$$\begin{cases} 6 - 15 = -9 \\ 9 - 9 = 0 \end{cases}$$

$$\begin{cases} -9 = -9 \\ 0 = 0 \end{cases}$$

Since our equations are still in balance, this means that our calculation was correct.

7. Fractions example

Solve for **x** and **y** :

$$\begin{cases} \dfrac{1}{2}x - \dfrac{2}{3}y = -1 \\[2em] \dfrac{3}{2}x + \dfrac{1}{3}y = 4 \end{cases}$$

Solution

First, let's see whether we can get rid of fractions so that we can calculate easier! We have **2** and **3** in the denominator in both equations. If we multiply both equations by the "Least Common Denominator", which is **6** , then we will remove all fractions:

$$\begin{cases} \dfrac{1}{2}x - \dfrac{2}{3}y = -1 \quad / \times 6 \\[2em] \dfrac{3}{2}x + \dfrac{1}{3}y = 4 \quad\;\; / \times 6 \end{cases}$$

$$\begin{cases} 6 \times \dfrac{1}{2}x - 6 \times \dfrac{2}{3}y = 6 \times (-1) \\[2em] 6 \times \dfrac{3}{2}x + 6 \times \dfrac{1}{3}y = 6 \times 4 \end{cases}$$

$$\begin{cases} 3x - 4y = -6 \\ 9x + 2y = 24 \end{cases}$$

Let's multiply the second equation by **2**. By doing so we will eliminate **y**:

21

$$\begin{cases} 3x - 4y = -6 \\ 9x + 2y = 24 \quad / \times 2 \end{cases}$$

$$\begin{cases} 3x - 4y = -6 \\ 18x + 4y = 48 \end{cases}$$

Add the equations side by side:

$$(3x - 4y) + (18x + 4y) = -6 + 48$$
$$3x - 4y + 18x + 4y = 42$$
$$21x = 42$$
$$x = \frac{42}{21}$$
$$x = 2$$

Now, substitute **x** in any equation with its value. Let's pick the simplest one:

$$3x - 4y = -6$$
$$3 \times 2 - 4y = -6$$
$$6 - 4y = -6$$
$$-4y = -6 - 6$$
$$-4y = -12$$
$$y = \frac{-12}{-4}$$
$$y = 3$$

We found the solutions for our System of Equations, and they are:

$$x = 2 \text{ and } y = 3$$

Check your work

Let's plug in the values of the variables in the equations and see whether they are still in balance.

Our System of Equations is:

$$\begin{cases} \dfrac{1}{2}x - \dfrac{2}{3}y = -1 \\[3mm] \dfrac{3}{2}x + \dfrac{1}{3}y = 4 \end{cases}$$

After we plug in the values:

$$\begin{cases} \dfrac{1}{2} \times 2 - \dfrac{2}{3} \times 3 = -1 \\[3mm] \dfrac{3}{2} \times 2 + \dfrac{1}{3} \times 3 = 4 \end{cases}$$

$$\begin{cases} 1 - 2 = -1 \\ 3 + 1 = 4 \end{cases}$$

$$\begin{cases} -1 = -1 \\ 4 = 4 \end{cases}$$

This proves that our calculation was correct.

8. Square roots example

Solve for **x** and **y** :

$$\begin{cases} 3x - y = 2 \\ \sqrt{2}x + \dfrac{1}{\sqrt{2}}y = \sqrt{32} \end{cases}$$

Solution

Let's see whether we can get rid of the Radical so that we can calculate easier! We have $\sqrt{2}$ in the denominator and as a coefficient of **x** in the second equation. If we multiply the second equation by $\sqrt{2}$, we will be able to greatly simplify the equation:

$$\begin{cases} 3x - y = 2 \\ \sqrt{2}x + \dfrac{1}{\sqrt{2}}y = \sqrt{32} \qquad / \times \sqrt{2} \end{cases}$$

$$\begin{cases} 3x - y = 2 \\ \sqrt{2} \times \sqrt{2}x + \sqrt{2} \times \dfrac{1}{\sqrt{2}}y = \sqrt{2} \times \sqrt{32} \end{cases}$$

The denominator gets eliminated:

$$\begin{cases} 3x - y = 2 \\ \sqrt{2} \times \sqrt{2}x + y = \sqrt{2} \times \sqrt{32} \end{cases}$$

Notice the coefficient of **x**, and combining the radicals:

$$\begin{cases} 3x - y = 2 \\ \left(\sqrt{2}\right)^2 x + y = \sqrt{2 \times 32} \end{cases}$$

Cancel out $\sqrt{}$ and ^{2} in front of **x**:

$$\begin{cases} 3x - y = 2 \\ 2x + y = \sqrt{64} \end{cases}$$

$$\begin{cases} 3x - y = 2 \\ 2x + y = 8 \end{cases}$$

Since the coefficients of **y** have the same value but with opposite sign $(+, -)$, we can simply use the Addition method and add the equations:

$$(3x - y) + (2x + y) = 2 + 8$$

The variable **y** gets eliminated:

$$3x - y + 2x + y = 2 + 8$$

$$5x = 10$$

$$x = \frac{10}{5}$$

$$x = 2$$

Substitute **x** in any equation and solve for **y**. Let's pick the first one:

$$3x - y = 2$$

$$3 \times 2 - y = 2$$

$$6 - y = 2$$

$$-y = 2 - 6$$

$$-y = -4$$

Multiply both sides of the equation by $(-\mathbf{1})$:

$$y = 4$$

So, the solutions for our System of Equations are:

$$x = 2 \quad \text{and} \quad y = 4$$

Check your work

Let's plug in the values of the variables in the equations and see whether they are still in balance.

Our simplified System of Equations is:

$$\begin{cases} 3x - y = 2 \\ 2x + y = 8 \end{cases}$$

After we plug in the values for **x** and **y**:

$$\begin{cases} 3 \times 2 - 4 = 2 \\ 2 \times 2 + 4 = 8 \end{cases}$$

$$\begin{cases} 2 = 2 \\ 8 = 8 \end{cases}$$

This proves that our calculation was correct.

9. Reinforcing Addition Method

Solve for **x** and **y** :

$$\begin{cases} 3x + 4y = 25 \\ 4x + 3y = 24 \end{cases}$$

Solution

Let's use the Addition method and try to eliminate **y**. Multiply the first equation by (-3) and the second equation by **4**:

$$\begin{cases} 3x + 4y = 25 & / \times (-3) \\ 4x + 3y = 24 & / \times 4 \end{cases}$$

$$\begin{cases} (-3) \times (3x + 4y) = (-3) \times 25 \\ \quad 4 \times (4x + 3y) = 4 \times 24 \end{cases}$$

$$\begin{cases} -9x - 12y = -75 \\ 16x + 12y = 96 \end{cases}$$

Add the equations side by side:

$$(-9x - 12y) + (16x + 12y) = -75 + 96$$

$$-9x - 12y + 16x + 12y = 21$$

$$7x = 21$$

$$x = \frac{21}{7}$$

$$x = 3$$

Substitute **x** in any equation and solve for **y**. Let's pick the first one:

$$3x + 4y = 25$$

$$3 \times 3 + 4y = 25$$

$$9 + 4y = 25$$

$$4y = 25 - 9$$

$$4y = 16$$

$$y = 4$$

So, the solutions for our System of Equations are:

$$x = 3 \quad \text{and} \quad y = 4$$

Check your work

Let's plug in the values of the variables in the equations and see whether they are still in balance.

Our System of Equations is:

$$\begin{cases} 3x + 4y = 25 \\ 4x + 3y = 24 \end{cases}$$

After we plug in the values for **x** and **y**:

$$\begin{cases} 3 \times 3 + 4 \times 4 = 25 \\ 4 \times 3 + 3 \times 4 = 24 \end{cases}$$

$$\begin{cases} 25 = 25 \\ 24 = 24 \end{cases}$$

This proves that our calculation was correct.

10. Reinforcing Substitution Method

Solve for **x** and **y** using Substitution method:

$$\begin{cases} 3x - 4y = 1 \\ 4x - 3y = 6 \end{cases}$$

Solution

Let's pick the first equation and solve for **x:**

$$3x - 4y = 1$$
$$3x = 1 + 4y$$
$$x = \frac{1 + 4y}{3}$$

Substitute **x** in the second equation:

$$4x - 3y = 6$$

After substitution:

$$4 \times \frac{1 + 4y}{3} - 3y = 6$$

$$\frac{4 \times (1 + 4y)}{3} - 3y = 6$$

$$\frac{4 + 16y}{3} - 3y = 6$$

Multiply both sides by **3**:

$$\frac{4 + 16y}{3} - 3y = 6 \qquad / \times 3$$

$$3 \times \left(\frac{4 + 16y}{3} - 3y\right) = 3 \times 6$$

$$3 \times \frac{4 + 16y}{3} - 3 \times 3y = 18$$

Denominator cancels out:

$$4 + 16y - 9y = 18$$

$$7y = 18 - 4$$

$$7y = 14$$

$$y = \frac{14}{7}$$

$$y = 2$$

Substitute the value of **y** in the first equation and solve for **x**. Let's pick the one where we isolated **x**:

$$x = \frac{1 + 4y}{3}$$

$$x = \frac{1 + 4 \times 2}{3}$$

$$x = \frac{1 + 8}{3}$$

$$x = \frac{9}{3}$$

$$x = 3$$

So, we found that the solutions for our System of Equations are:

$$x = 3 \quad \text{and} \quad y = 2$$

Check your work

Let's plug in the values of the variables in the equations and see whether they are still in balance.

Our System of Equations is:

$$\begin{cases} 3x - 4y = 1 \\ 4x - 3y = 6 \end{cases}$$

After we plug in the values for **x** and **y**:

$$\begin{cases} 3 \times 3 - 4 \times 2 = 1 \\ 4 \times 3 - 3 \times 2 = 6 \end{cases}$$

$$\begin{cases} 1 = 1 \\ 6 = 6 \end{cases}$$

This proves that our calculation was correct.

11. Introducing the Graph method

Solve for **x** and **y** using the **Graph method**:

$$\begin{cases} x + y = 3 \\ x - y = 1 \end{cases}$$

Solution

In the previous examples, we used two methods for solving the Systems of Equations: **"Substitution Method"** and the **"Addition Method"**. Another way to find the solutions of a System of Equations is the **"Graph Method"**.

It is called "Graph method" because we draw both equations on a coordinate plane, and the point $(\mathbf{x}, \mathbf{y})$ where the lines will intersect will represent the solutions for **x** and **y**.

In order to draw the equations, we will isolate first **y**, and then give arbitrary values to **x** so that we can find the value of **y** and plot them on the coordinate plane.

Let's isolate **y** in both equations:

$$\begin{cases} x + y = 3 \\ x - y = 1 \end{cases}$$

$$\begin{cases} y = 3 - x \\ -y = 1 - x \end{cases}$$

$$\begin{cases} y = 3 - x \\ y = x - 1 \end{cases}$$

Let's graph the first equation: $\mathbf{y = 3 - x}$. First we need to find at least two points $(\mathbf{x}, \mathbf{y})$ and then draw a line through those points. We will give arbitrary values to **x** and find **y**:

So, if $x = 1$ then:

$$y = 3 - x$$
$$y = 3 - 1$$
$$y = 2$$

We just found a point (x, y) on the line $y = 3 - x$. The point is $(1, 2)$ and let's call it **A**.

If $x = 3$ then:

$$y = 3 - x$$
$$y = 3 - 3$$
$$y = 0$$

We just found a second point (x, y) on the line $y = 3 - x$. The Point has the coordinates $(3, 0)$ and let's call it **B**.

Now we are going to draw the points $A = (1, 2)$ and $B = (3, 0)$ on a coordinate plane and draw a line through them:

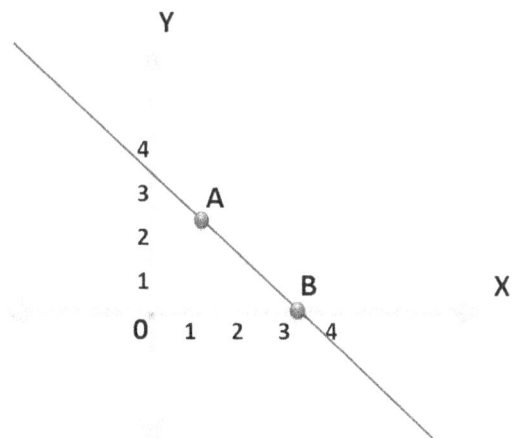

Now let's graph the second equation: $y = x - 1$. First we need to find at least two points (x, y) and then draw a line through those points. We will give arbitrary values to **x** and find **y**:

So, if **x = 1** then:

$$y = x - 1$$
$$y = 1 - 1$$
$$y = 0$$

We just found a point $(\mathbf{x}, \mathbf{y})$ on the line **y = x − 1**. The Point is $(\mathbf{1}, \mathbf{0})$ and let's call it **C**.

If **x = 3** then:

$$y = x - 1$$
$$y = 3 - 1$$
$$y = 2$$

We just found a second point $(\mathbf{x}, \mathbf{y})$ on the line **y = x − 1**. The point is $(\mathbf{3}, \mathbf{2})$ and let's call it **D**.

Now we are going to draw the points **C** = $(\mathbf{1}, \mathbf{0})$ and **D** = $(\mathbf{3}, \mathbf{2})$ on a coordinate plane and draw a line through them:

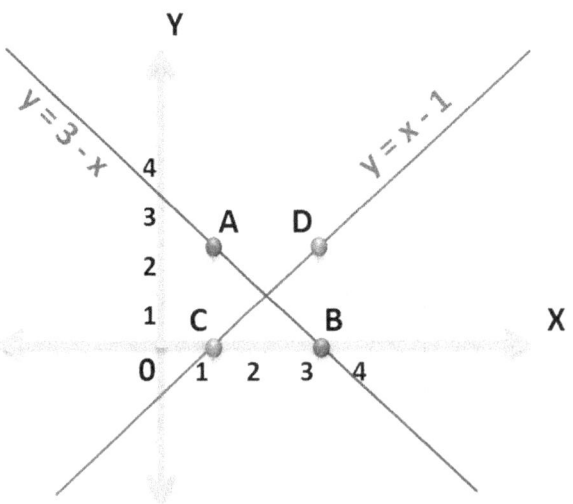

As you can see, the two lines or the two equations do cross each other. This intersection Point is what is common between them, and exactly this Point represents the solution for our system of equations.

We can see the Point on the graph below that its coordinates are $(\mathbf{2}, \mathbf{1})$.

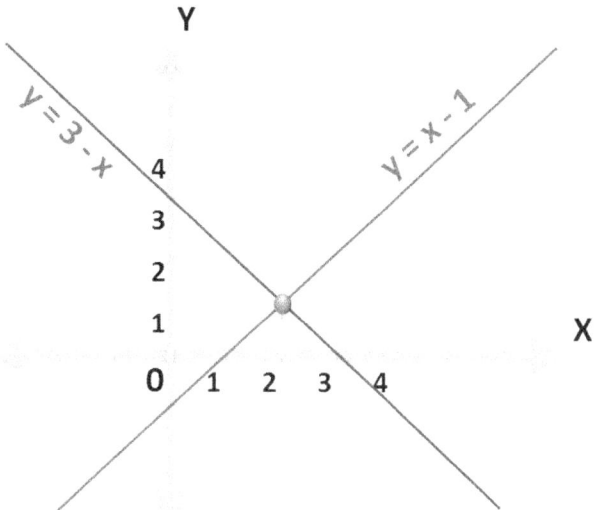

In fact, we did solve the same system of equations in Example 2. There we used the **"Addition Method"** and found **x = 2** and **y = 1**, which are the same values as the ones we found using the Graph method.

Check your work

Let's plug in the values of the variables in the equations and see whether they are still in balance.

Our System of Equations is:
$$\begin{cases} x + y = 3 \\ x - y = 1 \end{cases}$$
After we plug in the values for **x** and **y**:
$$\begin{cases} 2 + 1 = 3 \\ 2 - 1 = 1 \end{cases}$$
$$\begin{cases} 3 = 3 \\ 1 = 1 \end{cases}$$

Since our equations are still in balance, this means that our result was correct.

12. Reinforcing Graph method

Solve for **x** and **y** using the Graph Method:

$$\begin{cases} x + y = 4 \\ 2x + 2y = 8 \end{cases}$$

Solution

In the previous example, we used the **"Graph Method"** to solve a System of Equations to demonstrate the third method, in addition to the **"Substitution Method"** and the **"Addition Method"**, as a way to find the solutions of a System of Equations.

Again, it is called "Graph method" because we draw both equations on a coordinate plane, and the point $(\mathbf{x}, \mathbf{y})$ where the lines intersect will represent the solution for **x** and **y**.

In order to draw the equations, we give arbitrary values to **x** so that we can find the value of **y**, and plot them on the coordinate plane.

Let's give values to **x** in the first equation:

If **x = 3** then:

$$x + y = 4$$
$$3 + y = 4$$
$$y = 4 - 3$$
$$y = 1$$

We just found one point $(\mathbf{3}, \mathbf{1})$. Let's find another point:

If $x = 1$ then:

$$x + y = 4$$
$$1 + y = 4$$
$$y = 4 - 1$$
$$y = 3$$

We have now a second point $(1, 3)$.

Let's draw the points **B** $(3, 1)$ and **A** $(1, 3)$ on a coordinate plane and draw a line through them:

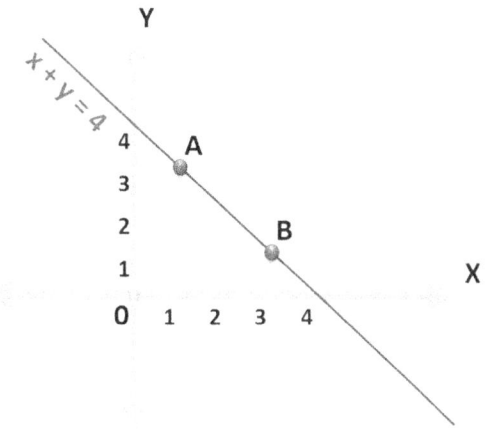

Now, let's graph the second equation: $2x + 2y = 8$. We need to find at least two points (x, y) and then draw a line through those points. Again, we will give arbitrary values to **x** and find the value of **y**:

So, if $x = 0$ then:

$$2x + 2y = 8$$
$$2 \times 0 + 2y = 8$$
$$2y = 8$$
$$y = \frac{8}{2}$$
$$y = 4$$

We found one point $(0, 4)$. Let's find another point:

If $x = 4$ then:

$$2x + 2y = 8$$
$$2 \times 4 + 2y = 8$$
$$8 + 2y = 8$$
$$2y = 8 - 8$$
$$2y = 0$$
$$y = 0$$

We have now a second point $(\mathbf{4, 0})$.

Let's draw the points $\boldsymbol{C}\ (\mathbf{0, 4})$ and $\mathbf{D}\ (\mathbf{4, 0})$ on the same coordinate plane where we graphed the first equation $\mathbf{x + y = 4}$:

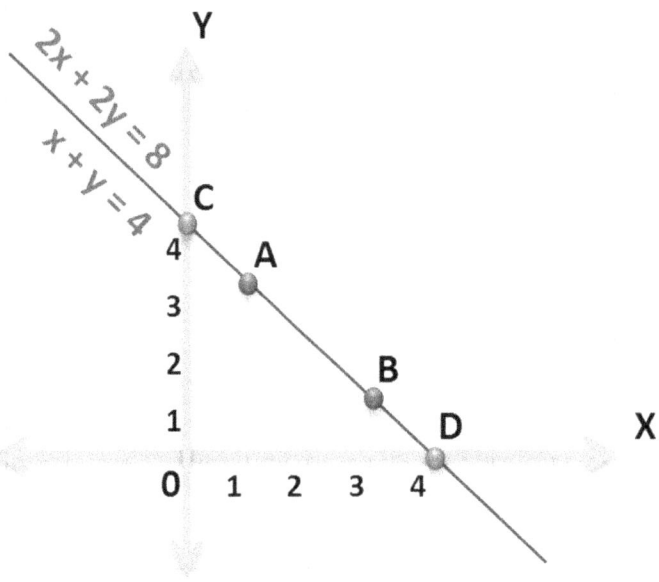

As you can see, the two Points are on the same line, the line of the first equation. We were expecting to get a new line and find the point where the two lines will intersect and this Point would have been our solution for the System of Equations, but we got only one line. Can you guess why we did not get a second line?

This is because we did not really have a System of Equations, but something that looks like System of Equations. If you take a closer look at the two equations, you will find that we have one and only one equation with two variables. The second equation is in fact the modified first equation. If we simplify the second equation we will get the first one:

Let's divide all terms of the second equation by **2**

$$2x + 2y = 8 \quad / \div 2$$

$$\frac{2x}{2} + \frac{2y}{2} = \frac{8}{2}$$

$$x + y = 4$$

What do you do in this case?

Well, from a single equation with two variables one cannot solve for a variable. What you can do is assign arbitrary value to one variable to calculate the other. This way you can find many combinations of (x, y) Points. In Math terminology we say, the equation with two variables has infinite solutions. In English, the line has unlimited Points.

13. The sum and difference of two numbers

The sum of two numbers is **13** and their difference is **1**.

Find the two numbers?

Solution

This type of word problems can be thought to be the simplest word problem that can be solved using the System of Equations.

Whenever we solve word problems, the **first** important step would be to really **understand** the problem or the situation. We do this by clearly identifying and understanding the facts that are given in the question.

From the wording of our question, we understand that there are three facts given to us:

1. "Two numbers" is one fact, which indicates that what we are looking for are two numbers which must satisfy some conditions.
2. "The sum is **13**". This is the second given fact. This is one of the conditions that those "two numbers" must satisfy.
3. "Their difference is **1**". This is the third given fact. This is another of the conditions that those "two numbers" must satisfy.

The **second** important step when we solve "word problems" would be to write or "translate" the facts given in sentences into a mathematical notation. Let's try this for the three facts we identified above:

1. "Two numbers". Since these two numbers are unknown to us at this moment, we will simply use two Math symbols for unknowns such as **x** and **y**.
2. "The sum is **13**". This means that when we add those two unknowns, the result will be **13**. In Math notation we write this like:

$$x + y = 13$$

3. "Their difference is **1**". This means that when we subtract one unknown from the other, the result will be **1**. In math notation we write:

$$x - y = 1$$

The **third** important step, when we solve "word problems", is to construct a Mathematical model and solve it. Since both conditions above must be satisfied for the two unknowns **x** and **y**, then the two equations make a System of Equations, and we write them like:

$$\begin{cases} x + y = 13 \\ x - y = 1 \end{cases}$$

This System of Equations is our Mathematical model, which we need to solve and find the solutions. Let's solve it then.

We can use either the **"Substitution Method"** or the **"Addition Method"** for solving System of Equations. Let's solve our system using the **"Addition Method"**, because it looks like we can quickly eliminate the **y** variable:

$$\begin{cases} x + y = 13 \\ x - y = 1 \end{cases}$$

Add the sides:

$$(x + y) + (x - y) = 13 + 1$$

Remove parenthesis:

$$x + y + x - y = 13 + 1$$

Add the "like" terms:

$$2x = 14$$

Variable **y** was eliminated, and now we can find **x**:

$$x = \frac{14}{2}$$

$$x = 7$$

Substitute **x = 7** in any of the equations to find **y**. Let's pick the first equation:

$$x + y = 13$$

$$y = 13 - x$$

$$y = 13 - 7$$

$$y = 6$$

So, we found that the solutions for our System of Equations are **x = 7** and **y = 6**.

Check your work

Let's use **x = 7** and **y = 6** and confirm that our equations are in balance.

Our System of Equations is:

$$\begin{cases} x + y = 13 \\ x - y = 1 \end{cases}$$

After we plug in the values for **x** and **y**:

$$\begin{cases} 7 + 6 = 13 \\ 7 - 6 = 1 \end{cases}$$

$$\begin{cases} 13 = 13 \\ 1 = 1 \end{cases}$$

This proves that our calculations were correct.

14. The sum and product of two numbers

The sum of two numbers is **9** and their product is **18**.

What are the two numbers?

Solution

This type of word problems can be thought to be the simplest word problem that can be solved using the System of Equations. Even though you might guess the two numbers in this easy example with small numbers, but as you will see, modeling the word problem and solving it mathematically can get a little bit more involved, and this the real benefit of this example.

Again, whenever we solve word problems, the **first** important step would be to really **understand** the problem or the scenario. We do this by clearly understanding the facts that are given.

From the wording of our question, we understand that there are three facts given to us:

1. "Two numbers" is one fact, which indicates that what we are looking for are two numbers which must satisfy some conditions.
2. "The sum is **9**". This is a second fact given. This is one of the conditions that those "two numbers" must satisfy.
3. "Their product is **18**". This is the third given fact. This is another of the conditions that those "two numbers" must satisfy.

The **second** important step, when we solve "word problems", would be to write or "translate" the given facts in sentences into a mathematical notation. Let's try this for the three facts we identified above:

1. "Two numbers". Since these two numbers are unknown to us at this moment, we will simply use two math symbols for unknowns such as **x** and **y** to mark the unknowns.

2. "The sum is **9**". This means that when we add those two unknowns, the result will be 9. In math notation we write:

$$x + y = 9$$

3. "Their product is **18**". This means that when we multiply the two unknowns, the result will be **18**. In math notation we can write this fact like:

$$xy = 18$$

Since both conditions must be satisfied for the two unknowns **x** and **y**, then the two equations make a System of Equations, and we write them like:

$$\begin{cases} x + y = 9 \\ xy = 18 \end{cases}$$

The **third** important step, when we solve "word problems", is to construct the Mathematical model and solve it. In our case, the model is a System of Equations. Let's solve it then.

We can use either the **"Substitution Method"** or the **"Addition Method"** for solving System of Equations.

Let's solve our system using the **"Substitution Method"**:

$$\begin{cases} x + y = 9 \\ xy = 18 \end{cases}$$

One way is to isolate **x** in the first equation:
$$x + y = 9$$
$$x = 9 - y$$

Substitute the value of **x**, which is **9 − y**, in the second equation:

$$xy = 18$$
$$(9 - y)y = 18$$

Remove the parenthesis and isolate **y**:

$$9y - y^2 = 18$$

We got a quadratic equation. Let's bring it in a proper quadratic form:

$$-y^2 + 9y - 18 = 0$$

How do we solve quadratic equations?

Recall first the general form of the quadratic equation:

$$ay^2 + by + c = 0$$

The formula to find the solutions of a quadratic equation is:

$$y_{1,2} = \frac{-b \pm \sqrt{b^2 - 4ac}}{2a}$$

In our equation we have:

$$a = -1, \qquad b = 9, \qquad c = -18$$

Let's substitute these values in the formula. So we have:

$$y_{1,2} = \frac{-b \pm \sqrt{b^2 - 4ac}}{2a}$$

$$= \frac{-9 \pm \sqrt{9^2 - 4 \times (-1) \times (-18)}}{2 \times (-1)}$$

$$= \frac{-9 \pm \sqrt{81 - 72}}{-2}$$

$$= \frac{-9 \pm \sqrt{9}}{-2}$$

$$= \frac{-9 \pm 3}{-2}$$

Now we can find the two solutions y_1 and y_2 of our quadratic equation:

$$y_1 = \frac{-9 + 3}{-2} = \frac{-6}{-2} = 3$$

$$y_2 = \frac{-9 - 3}{-2} = \frac{-12}{-2} = 6$$

Since quadratic equations have two solutions, and we got two solutions for y, then we will use one y at a time and find the corresponding x.

Let's use $y_1 = 3$ to find x using the first equation:

$$x + y = 9$$
$$x + 3 = 9$$
$$x = 9 - 3$$
$$x = 6$$

So, one solution for our system of equations is $x = 6$ and $y = 3$.

Let's use now the second solution $y_2 = 6$ to find x using the first equation:

$$x + y = 9$$
$$x + 6 = 9$$
$$x = 9 - 6$$
$$x = 3$$

So, the second solution for our system of equations is $x = 3$ and $y = 6$.

Notice that we don't really have two solutions for our system of equations, even though it appears so. In fact, if you take a closer look, the two numbers we were looking for are **3** and **6**.

Check your work

Let's use **x = 3** and **y = 6** and confirm that our equations are in balance.

Our System of Equations is:

$$\begin{cases} x + y = 9 \\ xy = 18 \end{cases}$$

After we plug in the values for **x** and **y**:

$$\begin{cases} 3 + 6 = 9 \\ 3 \times 6 = 18 \end{cases}$$

$$\begin{cases} 9 = 9 \\ 18 = 18 \end{cases}$$

This proves that our calculations were correct.

15. The relationship between two numbers

The sum of two numbers is **27** and one number is half the other.

What are the two numbers?

Solution

Let's start with the first step of solving the Systems of Equations, which is to understand the problem through the given facts.

The following three facts are obvious from the wording of the question:

1. "Two numbers" is a fact, which indicates that we are looking for two numbers that must satisfy some conditions.
2. "The sum is **27**" is the second fact, which tells us a condition that those "two numbers" must satisfy.
3. "One number is half the other" is the third fact that tells us about a certain relationship between those "two numbers".

The second important step, when we solve "word problems", would be to write or "translate" the facts given in words into a Mathematical notation. Let's try this for the three facts we identified above:

1. "Two numbers" are the two unknowns or variables and we will use **x** and **y** to mark them.
2. "The sum is **27**", which says that when we add those two variables, the result will be **27**. In math notation we write this like:

$$x + y = 27$$

3. "One number is half the other". This seems to be tricky, but it is not. We know we have two unknowns **x** and **y**, but we cannot tell if **x** is half of **y** OR **y** is half of **x**. The answer to this is that either statement is true because they are still unknowns and we

are in the process of assigning letters to numbers. At this stage it does not matter whether we name one number **x** or **y**. We have to decide and pick one, and once we decide, we have to stick with it. Let's pick the second one and say that **y** is half of **x**. In math notation we write:

$$y = \frac{x}{2}$$

In case this is still not clear, if **y** is to be half of **x**, then we need to split **x** in half and we do this be dividing **x** by **2**.

Since both conditions must be satisfied for the two unknowns **x** and **y**, then the two equations make a System of Equations, and we write them like:

$$\begin{cases} x + y = 27 \\ y = \dfrac{x}{2} \end{cases}$$

Now we need to execute the third step, which is to solve our Mathematical model or the System of Equations. Let's solve our system of equations using the **"Substitution Method"**.

Since **y** is already isolated in the second equation, we can just substitute it in the first equation and solve for **x**:

$$x + y = 27$$

$$x + \frac{x}{2} = 27$$

Multiply every term by **2** to eliminate the denominator:

$$x + \frac{x}{2} = 27 \qquad / \times 2$$

$$2 \times x + 2 \times \frac{x}{2} = 2 \times 27$$

$$2x + x = 54$$

$$3x = 54$$

$$x = \frac{54}{3}$$

$$x = 18$$

Now that we found **x**, we can use it to find **y**. Let's use the second equation:

$$y = \frac{x}{2}$$

$$y = \frac{18}{2}$$

$$y = 9$$

So, we found that the solutions for our system of equations are **x = 18** and **y = 9**.

Check your work

Let's use **x = 18** and **y = 9** and confirm that our equations are in balance.

Our System of Equations is:

$$\begin{cases} x + y = 27 \\ y = \dfrac{x}{2} \end{cases}$$

After we plug in the values for **x** and **y**:

$$\begin{cases} 18 + 9 = 27 \\ 9 = \dfrac{18}{2} \end{cases}$$

$$\begin{cases} 27 = 27 \\ 9 = 9 \end{cases}$$

This proves that our calculations were correct.

16. Chocolates and Candies

Nina and Artie are selling Chocolates and Candies and are planning to donate the money to their school basketball team to help them buy the uniforms.

Nina sold **80** Chocolates and **40** Candies and was able to donate **$200** to the basketball team. Artie sold **36** Chocolates and **75** Candies and was able to contribute **$147** to the basketball team.

If Nina and Artie charged the same price for Chocolates and Candies, then how much did they charge for one Chocolate and how much for one Candy?

Solution

As usual, we will start with the first step and clearly understand the facts that are given.

From the wording of our question, we understand that there are three facts given to us:

1. "Chocolates and Candies" is one fact. These two items are being sold and we need to find the price of each.
2. "Nina's sales are **$200**". In other words, a combination of **80** Chocolates and **40** Candies totaled **$200**.
3. "Artie's sales are **$147**". This is the third fact and it means that a combination of **36** Chocolates and **75** Candies sold totaled **$147**.

Our second important step, when we solve "word problems", would be to write or "translate" the facts given in sentences into a mathematical notation. Let's try this for the three facts we identified above:

1. "Chocolates and Candies". Since we need to find the price of each item AND the two prices are unknown to us at this moment, we will simply use two math symbols for unknowns such as **x** and **y**. Let's use **x** for the price of a Chocolate and **y** for the price of a Candy.

2. "Nina's sales are **$200** and she sold **80** Chocolates and **40** Candies". If we use **x** and **y**, then this fact in Math notation can be written like:

$$80x + 40y = 200$$

3. "Artie's sales are **$147** and he sold **36** Chocolates and **75** Candies". If we use **x** and **y**, then this fact can in math notation can be written like:

$$36x + 75y = 147$$

Another important fact that should not be ignored is that both, Nina and Artie charged the same price for a chocolate and for a candy, which means that the **x** and **y** in the first equation are the same **x** and **y** of the second equation. This fact tells us that the two equations above are linked or are related and make a System of Equations (Note! If they would have charged different prices, then we would have not been able to find out the prices for a Chocolate and Candy based on the given facts). So, the following is our system of equations:

$$\begin{cases} 80x + 40y = 200 \\ 36x + 75y = 147 \end{cases}$$

The third step is to solve the Mathematical model or the System of Equations we just constructed. Let's solve it then.

We can use either the **"Substitution Method"** or the **"Addition Method"** for solving System of Equations. Let's solve our system using the **"Substitution Method"**:

$$\begin{cases} 80x + 40y = 200 \\ 36x + 75y = 147 \end{cases}$$

Let's isolate **y** in the first equation. It would be easier to calculate if we can make the numbers smaller. It should be obvious that if we divide all terms in the first equation by **40**, we will get an equivalent equation which is easier to work with:

$$\begin{cases} 80x + 40y = 200 \quad / \div 40 \\ 36x + 75y = 147 \end{cases}$$

$$\begin{cases} 2x + y = 5 \\ 36x + 75y = 147 \end{cases}$$

$$\begin{cases} y = 5 - 2x \\ 36x + 75y = 147 \end{cases}$$

Now we will substitute the first equation in the second equation and solve for **x**:

$$36x + 75y = 147$$
$$36x + 75(5 - 2x) = 147$$
$$36x + 375 - 150x = 147$$
$$36x - 150x = 147 - 375$$
$$-114x = -228$$
$$x = \frac{-228}{-114}$$
$$x = 2$$

Now that we found **x**, we can substitute its value in any equation and find out the value of **y**. Let's pick the second equation:

$$36x + 75y = 147$$
$$36 \times 2 + 75y = 147$$
$$72 + 75y = 147$$
$$75y = 147 - 72$$
$$75y = 75$$
$$y = \frac{75}{75}$$
$$y = 1$$

So, we found that the solutions for our system of equations are $x = 2$ and $y = 1$. Recall that we decided to use x for the price of a Chocolate and y for the price of a Candy. Hence, the price of a Chocolate is $2 and the price of a Candy is $1.

Check your work

Let's use $x = 2$ and $y = 1$ and confirm that our equations are in balance.

Our System of Equations is:

$$\begin{cases} 80x + 40y = 200 \\ 36x + 75y = 147 \end{cases}$$

After we plug in the values for x and y:

$$\begin{cases} 80 \times 2 + 40 \times 1 = 200 \\ 36 \times 2 + 75 \times 1 = 147 \end{cases}$$

$$\begin{cases} 200 = 200 \\ 147 = 147 \end{cases}$$

This proves that our calculations were correct.

17. Age difference

Let's say that Nina is **7** times older than her brother Artie. Five years from now, Nina will be **2** times older than Artie.

How old are Nina and Artie now?

Solution

Let's understand the facts that are given to us. From the wording of our question, we understand that there are three facts given to us:

1. "Nina's and Artie's ages" is one fact. Their ages are discussed, the relationship of their ages is described and we need to find out how old they are.
2. "Nina is **7** times older than Artie". This is another fact which describes one relationship between their ages.
3. "Five years from now, Nina will be **2** times older then Artie". This is the third fact that describes another, future relationship between their ages.

Let's "translate" the facts given in sentences into a mathematical notation:

1. "Nina's and Artie's ages". Since we need to find their ages AND they are unknown to us at this moment, we will simply use two math symbols for unknowns such as **x** and **y**. Let's use **x** for Nina's age, and **y** for Artie's age.
2. "Nina is **7** times older than Artie". If we use **x** and **y** then this fact in math notation can be written like:

$$x = 7y$$

3. "Five years from now, Nina will be **2** times older then Artie". The expression "Five years from now" is a critical one and must be properly understood. If Nina is now **x** years old, then "Five years from now" she will be **x + 5** years old. Similarly, if Artie is now **y** years old, then "Five years from now" he will be **y + 5** years old. If we

use **x** and **y** then the fact "Five years from now, Nina will be 2 times older then Artie" in math notation can be written like:

$$x + 5 = 2(y + 5)$$

Since the above facts are related, then we can use Systems of Equations to find out their ages. So, the following is our system of equations:

$$\begin{cases} x = 7y \\ x + 5 = 2(y + 5) \end{cases}$$

Let's solve our System of Equations.

We will use the **"Substitution Method"** for solving our System of Equations because **x** in the first equation is already isolated. Before we do the substitution, it is better to normalize or bring the second equation in a better shape:

$$\begin{cases} x = 7y \\ x + 5 = 2(y + 5) \end{cases}$$

$$\begin{cases} x = 7y \\ x + 5 = 2y + 10 \end{cases}$$

$$\begin{cases} x = 7y \\ x - 2y = 10 - 5 \end{cases}$$

$$\begin{cases} x = 7y \\ x - 2y = 5 \end{cases}$$

Now we will substitute the first equation in the second equation and solve for **x**:

$$x - 2y = 5$$
$$7y - 2y = 5$$
$$5y = 5$$

$$y = \frac{5}{5}$$

$$y = 1$$

Let's use the first equation to find the value of **x**:

$$x = 7y$$

Let's substitute the value of y:

$$x = 7 \times 1$$
$$x = 7$$

Since we used **x** for Nina's age and **y** for Artie's age, then **x = 7** means that Nina is now **7** years old and **y = 1** means that Artie is now **1** year old.

Check your work

Let's use **x = 7** and **y = 1** and confirm that our equations are in balance.

Our System of Equations is:

$$\begin{cases} x = 7y \\ x + 5 = 2(y + 5) \end{cases}$$

After we plug in the values for **x** and **y**:

$$\begin{cases} 7 = 7 \times 1 \\ 7 + 5 = 2(1 + 5) \end{cases}$$
$$\begin{cases} 7 = 7 \\ 12 = 12 \end{cases}$$

This proves that our calculations were correct.

18. Restaurant TIPS

Let's say that a friend of yours, who is in a Culinary School, found a summer Internship to work as a waiter in fine dining Restaurant. After few days you chat with him about the Internship and he tells you that the Food in that Restaurant is superb, the TIPS are good etc.(Note! TIPS are the money that waiters receive for providing good service to customers. TIPS is an acronym and it stands for: **T**o **I**nsure **P**roper **S**ervice.)

When you asked him how much TIPS he earns on average, he did not tell you the numbers, but because you are a Mathematician, he told you the following challenging story:

Last night they were **5** Waiters and **3** Busboys and shared **$650**. The night before last night they were **3** Waiters and **2** Busboys, they shared **$400**, and waiter's and busboy's share was the same as last night. As in most places, the Waiter's share is double the Busboy's share.

How much money did your friend make each of those two nights working as an Intern in that Restaurant?

Write a Mathematical model and find out how much TIPS a Waiter and a Busboy made each night.

Solution

Let's analyze and understand the problem first.

The waiters and busboys collect TIPS and at the end of the night they share them according to some Ratio (i.e. **1: 2**). In the two nights the Ratio of how the waiters and busboys split the money did not change. Waiter's share remained double the Busboy's share. What changed was the total TIPS they made and their number in each of the two shifts. Hence we will not consider the ratio when we model the problem.

Let's mark with x the dollar amount a waiter got each night, and y the dollar amount what a busboy got each night. Let's express in Math notation the situation of the last night:

They were **5** Waiters and **3** Busboys and shared **$650**. This can be written as:

$$5x + 3y = \$650$$

The night before last night they were **3** Waiters and **2** Busboys and shared **$400**. This can be written as:

$$3x + 2y = \$400$$

Now the challenge is how to find **x** and **y**. The two equations above are related to each other because your friend said he made the same amount of money each night and the ratio how the money is split between waiters and busboys did not change.

Since the above equations are related we can treat them as system of Equations and solve for **x** and **y**:

$$\begin{cases} 5x + 3y = 650 \\ 3x + 2y = 400 \end{cases}$$

We can use either Addition method or Substitution method. The Addition method seems to be the easiest one for this System of Equations.

The Addition method is also called Elimination method because when we add the two equations we eliminate one of the two variables and solve for the one which is left. Let's eliminate variable **y** . In order to eliminate **y** we need modify the equations in a way so that we get the same coefficient in front of **y** but with opposite signs $(+, -)$.

How to modify the equations then? You need to remember this:
1. If you multiply the left and right side of an equation with the same number, the equation will still remain in balance. It will just look differently. It is still equivalent equation.
2. Use your imagination and creativity to find one or two numbers to make the coefficients in front of **y** the same value but with opposite signs.

If you take a closer look, you can see that if you multiply the first equation with the coefficient of the **y** in the second equation which is **2**, AND you multiply the second equation with the coefficient of **y** in the first equation which is **3**, then you will get **6** in front of the y in both equations. To get the opposite sign you will have to use a negative of one of the coefficients.

Let's demonstrate this in action. Let's multiply the first equation with **2** and the second equation with $-$ **3**:

$$\begin{cases} 5x + 3y = \$650 & /\times 2 \\ 3x + 2y = \$400 & /\times (-3) \end{cases}$$

$$\begin{cases} 2 \times (5x + 3y) = 2 \times \$650 \\ -3 \times (3x + 2y) = -3 \times \$400 \end{cases}$$

$$\begin{cases} 10x + 6y = \$1,300 \\ -9x - 6y = -\$1,200 \end{cases}$$

Now that you have the same value **6** but with opposite sign in front of y, you can add the left and right sides of both equations, and the result will be:

$$(10x + 6y) + (-9x - 6y) = \$1,300 + (-\$1,200)$$
$$10x + 6y - 9x - 6y = \$1,300 - \$1,200$$
$$x = \$100$$

As you can see, you eliminated y ($6y - 6y = 0$).

Now that you have x, substitute its value in any of the equations to find the value of y. Let's use the second equations because it has smaller numbers and it is easier to work with:

$$3x + 2y = \$400$$

Substitute x with $\$100$:

$$3 \times \$100 + 2y = \$400$$
$$\$300 + 2y = \$400$$
$$2y = \$400 - \$300$$
$$2y = \$100$$

$$y = \frac{\$100}{2}$$

$$y = \$50$$

Back to the waiters and busboys in the Restaurant, we used x to mark the dollar amount a Waiter got each night, and we used y to mark Busboy's share. Since you found $x = \$100$ and $y = \$50$, this means that your Waiter friend earned $\$100$ each night, and a Busboy earned $\$50$.

Check your work

Let's use $x = 100$ and $y = 50$ and confirm that our equations are in balance.

Our System of Equations is:

$$\begin{cases} 5x + 3y = 650 \\ 3x + 2y = 400 \end{cases}$$

After we plug in the values for x and y:

$$\begin{cases} 5 \times 100 + 3 \times 50 = 650 \\ 3 \times 100 + 2 \times 50 = 400 \end{cases}$$

$$\begin{cases} 650 = 650 \\ 400 = 400 \end{cases}$$

This proves that our calculations were correct.

19. Bikes

Imagine you own a business where you sell various types of bikes: Bicycles, single-rider Tricycles and two-rider tandem bicycles. Let's say that the number of all bikes in the store is **45**, and the number of wheels of all bikes is **105**, and the number of all pedals is **110**.

Find out how many Bicycles, Tricycles and Tandem Bicycles you have for sale in the store?

Solution

Let's understand the facts that are given to us.

From the wording of our question, we understand that there are three facts given to us:

1. "The number of all bikes is **45**". We don't know their number by type, and that is what we need to find out.
2. "The number of wheels of all bikes is **105**". This is another fact which is given to us. We know that bicycles and tandem bikes have **2** wheels each, and tricycles have **3** wheels.
3. "The number of all pedals is **110**". This is the third fact given to us explicitly. We know that bicycles and tricycles have **2** pedals each, and tandem bikes have **4** pedals (**2** for each rider).

Now, let's "translate" the facts given in sentences into a mathematical notation:

1. "The number of all bikes is **45**". Since we need to find out the types of bikes we have for sale AND they are unknown to us at this moment, we will simply use three math symbols for unknowns such as **x**, **y**, and **z** for each type. Let's use:
   - **x** for the number of bicycles,
   - **y** for the number of tricycles, and
   - **z** for the number of tandem bikes.

Since we know that their number together is **45**, then in math notation we can write this fact like:

$$x + y + z = 45$$

2. "The number of wheels of all bikes is **105**". Since bicycles and tandem bikes have **2** wheels each, and tricycles have **3** wheels, then when we add all of them together we have **105** wheels. In math notation, this fact can be written as:

$$2x + 3y + 2z = 105$$

3. "The number of all pedals is **110**". Since bicycles and tricycles have **2** pedals each, and tandem bikes have **4** pedals (**2** for each rider), then when we add all pedals together we have **110** pedals. In math notation, this fact can be written as:

$$2x + 2y + 4z = 110$$

Since the above facts are related, then we can use the Systems of Equations to find out how many of each type of bikes we have for sale in our store. So, the following is our system of equations:

$$\begin{cases} x + y + z = 45 \\ 2x + 3y + 2z = 105 \\ 2x + 2y + 4z = 110 \end{cases}$$

So, we have a System of three equations with three variables. How do we solve this? Well, the principle remains the same. Let's use the **"Substitution Method"** for solving the systems of equations:

First we will isolate, let's say, **x** in one equation and then substitute in some other equation. Let's isolate **x** in the first equation:

$$\begin{cases} x = 45 - y - z \\ 2x + 3y + 2z = 105 \\ 2x + 2y + 4z = 110 \end{cases}$$

Let's substitute the isolated **x**, let's say, in the third equation, which is:

$$2x + 2y + 4z = 110$$

$$2 \times (45 - y - z) + 2y + 4z = 110$$

$$90 - 2y - 2z + 2y + 4z = 110$$

Luckily for us, **y** gets eliminated when we add the like terms, and now we have:

$$-2z + 4z = 110 - 90$$

$$2z = 20$$

$$z = \frac{20}{2}$$

$$z = 10$$

So, we found the value of **z**. Let's substitute this in the equation where we isolated **x**:

$$x = 45 - y - z$$

$$x = 45 - y - 10$$

$$x = 35 - y$$

Our equation became even simpler, and **x** is now dependent only on **y**.

We can use the values of **x** and **z** and substitute them in the second equation and be able to find the value of **y**, (note that **x** in this moment does not have a numerical value, but it has an algebraic value which we can still use as a substitution for **x**):

$$2x + 3y + 2z = 105$$

$$2 \times (35 - y) + 3y + 2 \times 10 = 105$$

$$70 - 2y + 3y + 20 = 105$$

$$-2y + 3y = 105 - 20 - 70$$

$$y = 15$$

So, we found the value of **y**. Let's substitute this in the equation where we isolated **x**:

$$x = 35 - y$$

$$x = 35 - 15$$

$$x = 20$$

Now we have all the numerical values for all three variables. We found that we have **20** bicycles (**x = 20**), **15** tricycles (**y = 15**) and **10** tandem bikes (**z = 10**) for sale in our store.

Check your work

Let's use **x = 20** , **y = 15** and **z = 10** and confirm that our equations are in balance.

Our System of Equations is:

$$\begin{cases} x + y + z = 45 \\ 2x + 3y + 2z = 105 \\ 2x + 2y + 4z = 110 \end{cases}$$

After we plug in the values for **x** and **y**:

$$\begin{cases} 20 + 15 + 10 = 45 \\ 2 \times 20 + 3 \times 15 + 2 \times 10 = 105 \\ 2 \times 20 + 2 \times 15 + 4 \times 10 = 110 \end{cases}$$

$$\begin{cases} 45 = 45 \\ 40 + 45 + 20 = 105 \\ 40 + 30 + 40 = 110 \end{cases}$$

$$\begin{cases} 45 = 45 \\ 105 = 105 \\ 110 = 110 \end{cases}$$

This proves that our calculations were correct.

20. Apple devices

Imagine that your grandfather decided to buy some Apple Devices, such as iPad, iPhone and Apple Watch, for you and your two siblings.

Since these devices are expensive, he was calculating to determine how much money he can afford to spend. If he buys one of each device to all his three grandchildren, he would spend $\$4,200$.

Since he cannot afford to spend that much, he was considering to get only one iPad so that you all can share it, but each one of you will get an iPhone and Apple Watch. This combination would cost him $\$3,000$.

Another option he considered is to get everyone of you an Apple Watch, one iPad for you all to share and only one iPhone for you because you are special. This set of Apple devices would cost him $\$2,000$.

Find out how much is the price of the iPad, the iPhone and the Apple Watch?

Solution

Let's understand the facts that are given to us.

From the wording of our question, we understand that there are three facts given to us:

1. "One of each device to all his three grandchildren" will total $\$4,200$. We don't know the price of each electronic device, but based on this total, that is what we need to find out.
2. "One iPad, **3** iPhones and **3** Apple Watches" total $\$3,000$. This is another fact which is given to us.
3. "One iPad, one iPhone and **3** Apple Watches" totaling $\$2,000$. This is the third fact given to us explicitly.

Now, let's "translate" the facts given in sentences into a mathematical notation:

Since we need to find out the prices of each Apple device AND they are unknown to us at this moment, we will simply use three Math symbols for unknowns such as **x, y**, and **z** for each device. Let's use: **x** for the price of an iPad, **y** for the price of an iPhone, and **z** for the price of an Apple Watch.

1. The fact "one of each device to all his three grandchildren" means that he will buy **3** iPads, **3** iPhones and **3** Apple watches. All these devices will cost him $**4, 200**. In Math notation we can write this fact like:

$$3x + 3y + 3z = 4200$$

2. "**1** iPad, **3** iPhones and **3** Apple Watches totaling $**3, 000**". In math notation this fact can be written as:

$$x + 3y + 3z = 3000$$

3. "One iPad, one iPhone and **3** Apple watches". In an equation form this fact can be written like:

$$x + y + 3z = 2000$$

Since the above facts are related, then we can use the Systems of Equations to find out the price of each Apple device. So, the following is our System of Equations:

$$\begin{cases} 3x + 3y + 3z = 4200 \\ x + 3y + 3z = 3000 \\ x + y + 3z = 2000 \end{cases}$$

So, we have a System of three equations with three variables. How do we solve this? Well, the principle remains the same. Let's use the **"Substitution Method"** for solving systems of equation:

First we will isolate, let's say, **x** in one equation and then substitute in some other equation. Let's isolate **x** in the second equation:

$$\begin{cases} 3x + 3y + 3z = 4200 \\ x = 3000 - 3y - 3z \\ x + y + 3z = 2000 \end{cases}$$

Let's substitute the isolated **x**, let's say, in the third equation, which is:

$$x + y + 3z = 2000$$

Let's substitute **x** and solve:

$$(3000 - 3y - 3z) + y + 3z = 2000$$
$$3000 - 3y - 3z + y + 3z = 2000$$

Luckily for us, **z** gets eliminated when we add the like terms, and we have:

$$-3y + y = 2000 - 3000$$
$$-2y = -1000$$
$$y = \frac{-1000}{-2}$$
$$y = 500$$

So, we found the value of **y**. Let's substitute this in the equation where we isolated **x**:

$$x = 3000 - 3y - 3z$$
$$x = 3000 - 3 \times 500 - 3z$$
$$x = 3000 - 1500 - 3z$$
$$x = 1500 - 3z$$

Our equation became even simpler, and **x** is now dependent only on **z**.

We can use the values of **x** and **y** and substitute them in the first equation, which is:

$$3x + 3y + 3z = 4200$$

After substitution:

$$3 \times (1500 - 3z) + 3 \times 500 + 3z = 4200$$
$$4500 - 9z + 1500 + 3z = 4200$$

$$6000 - 6z = 4200$$
$$-6z = 4200 - 6000$$
$$-6z = -1800$$
$$z = \frac{-1800}{-6}$$
$$z = 300$$

So, we found the value of **z**. Let's substitute this in the equation where we isolated **x**:

$$x = 1500 - 3z$$
$$x = 1500 - 3 \times 300$$
$$x = 1500 - 900$$
$$x = 600$$

Now we have all the values for all three variables. We found that iPad costs **\$600 (x = 600)**, iPhone costs **\$500 (y = 500)** and the Apple Watch costs **\$300 (z = 300)**.

Check your work

Let's use **x = 600**, **y = 500** and **z = 300**, and confirm that our equations are in balance.

Our System of Equations is:

$$\begin{cases} 3x + 3y + 3z = 4200 \\ x + 3y + 3z = 3000 \\ x + y + 3z = 2000 \end{cases}$$

After we plug in the values for **x** and **y**:

$$\begin{cases} 3 \times 600 + 3 \times 500 + 3 \times 300 = 4200 \\ 600 + 3 \times 500 + 3 \times 300 = 3000 \\ 600 + 500 + 3 \times 300 = 2000 \end{cases}$$

$$\begin{cases} 1800 + 1500 + 900 = 4200 \\ 600 + 1500 + 900 = 3000 \\ 600 + 500 + 900 = 2000 \end{cases}$$

$$\begin{cases} 4200 = 4200 \\ 3000 = 3000 \\ 2000 = 2000 \end{cases}$$

This proves that our calculations were correct.

21. Mortgage and Property Taxes

Let's say that your family is looking to buy a house or a condo. In order to see which property your family can afford, it is crucial to find out how much will be the Monthly payments. Specifically, they want to know how much will be the monthly mortgage payments and how much the property taxes.

Finally your family found an affordable property and your payments will be $\$2,300$ a month for the next **30** years (we assume the property taxes will not change and will remain fixed).

If the monthly mortgage payment is for $\$700$ higher than the monthly property tax payment, then how much is the mortgage payment and how much are the property taxes for the property your family is about to buy?

Solution

Let's understand and extract the facts that are given to us .From the wording of our question, we understand that there are few details given to us:

1. "Monthly payments consist of mortgage and property taxes."
2. "Mortgage and property taxes" are $\$2,300$ a month.
3. "The length of the mortgage is **30** years", and
4. "Monthly Mortgage payment is for $\$700$ higher than the property tax payment".

Now, let's "translate" the facts given in sentences into a mathematical notation:

1. Since we need to find out the monthly mortgage payments and property taxes AND they are unknown to us at this moment, we will simply use two math symbols for unknowns such as **m** and **t** (we don't have to use always **x** and **y**). Let's use: **m** for the mortgage, and **t** for taxes.
2. The fact "mortgage and property taxes are $\$2,300$ a month" means that when mortgage and taxes are added, the result is $\$2,300$. In math notation we can write this fact like:

$$m + t = 2300$$

3. "The length of the mortgage is **30** years". This detail is given, but if you take a closer look it is not relevant for our purposes. It is true that the length of the mortgage is very important variable when we calculate the monthly mortgage payments, but the calculation is done when the property is purchased and remains fixed for the entire period of **30** years. Here we are not re-calculating the mortgage based on time, but we are trying to determine it based on the relationship with taxes and the total monthly payment. Hence this fact can be ignored.

4. "Mortgage payment is for **$700** higher than the property tax payments". In other words, taxes plus **$700** equals mortgage payment. In a Math equation form this fact can be written like:

$$m = t + 700$$

Since the above facts are related, then we can use the Systems of Equations to find out the monthly mortgage payments and monthly property taxes. So, the following is our system of equations:

$$\begin{cases} m + t = 2300 \\ m = t + 700 \end{cases}$$

Let's use the **"Substitution Method"** to solve this system of equations:

The variable **m** is already isolated and we will substitute in the other equation, which is:

$$m + t = 2300$$

Let's substitute **m**:

$$(t + 700) + t = 2300$$

Let's solve for **t**:

$$t + 700 + t = 2300$$

$$2t + 700 = 2300$$

$$2t = 2300 - 700$$

$$2t = 1600$$

$$t = \frac{1600}{2}$$

$$t = 800$$

So, we found the value of **t**. Let's substitute this in the equation where m is isolated:

$$m = t + 700$$

$$m = 800 + 700$$

$$m = 1500$$

We found that monthly mortgage payments are $\mathbf{\$1,500}$ and monthly property taxes are $\mathbf{\$800}$.

Check your work

Let's use **m = 150** and **t = 800** and confirm that our equations are in balance.

Our System of Equations is:

$$\begin{cases} m + t = 2300 \\ m = t + 700 \end{cases}$$

After we plug in the values for **x** and **y**:

$$\begin{cases} 1500 + 800 = 2300 \\ 1500 = 800 + 700 \end{cases}$$

$$\begin{cases} 2300 = 2300 \\ 1500 = 1500 \end{cases}$$

This proves that our calculations were correct.

22. Planning for a Vacation

Let's say that you are planning for a vacation in the tropics and are considering booking a Hotel plus Flight package.

If you were to buy **1** night Hotel plus Flight package, the total price would be **$600**.

If you were to buy **7** nights Hotel plus Flight package, the total price would be **$1,200**.

Given that the Hotel does not give special discounts if you stay longer, and the price of the return trip flight ticket does not change depending on how apart the flights are, then how much does **1** night stay in Hotel cost, and how much does the return trip flight ticket cost?

Solution

Let's understand and extract the facts that are given to us.

From the wording of our question, we understand that the following details are given to us:

1. We are booking "Hotel + Flight package"
2. " **1** Night in a Hotel plus Flight package would be **$600**".
3. " **7** Nights in a Hotel plus Flight package would be **$1,200**", and
4. "No special discounts from the Hotel or Airline".

Now, let's "translate" the facts given in sentences into a mathematical notation:

1. Since we need to find out the cost of **1** night staying in a Hotel and the price of a return Flight ticket AND they are unknown to us at this moment, we will simply use two Math symbols for unknowns such as **n** and **f** (we don't have to use always **x** and **y**). Let's use: **n** for the night in a Hotel, and **f** for flight ticket.
2. The fact " **1** night Hotel plus Flight package would be **$600**" means that when the price for **1** night hotel stay and the price for a flight ticket are added, the result is **$600**. In math notation we can write this fact like:

$$n + f = 600$$

3. " **7** Nights in a Hotel plus Flight package would be **$1,200**". In other words, **7** nights in a hotel plus the flight ticket will cost **$1,200**. In an equation form this fact can be written like:

$$7n + f = 1200$$

Note that the coefficient of **n** changes the longer we stay, but the coefficient of **f** stays **1** (we don't write it but it is understood) because we need only one return trip flight ticket. We don't fly every day during our vacation, but only once to get in the tropics and once to return home, no matter how long we stay there.

Since the above facts are related, then we can use the Systems of Equations to find out the cost of **1** night in a Hotel and the cost for return Flight Ticket. So, the following is our system of equations:

$$\begin{cases} n + f = 600 \\ 7n + f = 1200 \end{cases}$$

Let's use the **"Substitution Method"** to solve this system of equations. We will first isolate, let's say, **f** in the first equation:

$$\begin{cases} f = 600 - n \\ 7n + f = 1200 \end{cases}$$

We will now substitute **f** in the other equation, which is:

$$7n + f = 1200$$

Let's substitute **f**:

$$7n + (600 - n) = 1200$$

Let's solve for **n**:

$$7n + 600 - n = 1200$$
$$7n - n = 1200 - 600$$
$$6n = 600$$

$$n = \frac{600}{6}$$
$$n = 100$$

So, we found the value of **n**. Let's substitute this in the equation where **f** is isolated:

$$f = 600 - n$$
$$f = 600 - 100$$
$$f = 500$$

We found that **1** night in a Hotel will cost **$100** and the Flight Ticket will cost **$500**.

Check your work

Let's use **n = 100** and **f = 500** and confirm that our equations are in balance.

Our System of Equations is:

$$\begin{cases} n + f = 600 \\ 7n + f = 1200 \end{cases}$$

After we plug in the values for **x** and **y**:

$$\begin{cases} 100 + 500 = 600 \\ 7 \times 100 + 500 = 1200 \end{cases}$$

$$\begin{cases} 600 = 600 \\ 1200 = 1200 \end{cases}$$

This proves that our calculations were correct.

23. How many questions on a test

Let's say that your next Math test will have **15** questions worth **100** points. The test is all about Systems of Equations and the questions are of two types. Most of the questions are Systems of Equations with **two variables and two equations** and these are worth **5** points each. The rest of the questions are Systems of Equations with **three variables and three equations** which are worth **10** points each.

Find out how many questions of each type of Systems of Equations are on the test?

Solution

Let's understand and extract the facts that are given to us.

From the wording of our question, we understand that there many details given to us:

1. "Two types of questions,"
2. **15** questions in total,
3. The "two equations" type is worth **5** points,
4. The "three equations" type is worth **10** points, and
5. "Total points on the test are **100**".

Now, let's "translate" the facts given in sentences into a mathematical notation:

1. Since we need to find out how many of each type questions are on the test AND they are unknown to us at this moment, we will simply use two math symbols for unknowns such as **x** and **y**. Let's use:
   - **x** for the number of Systems of Equations with **2** variables and **2** equations, and
   - **y** for the number of Systems of Equations with **3** variables and **3** equations.
2. The fact that there are **15** questions in total, in math notation can be written like:

$$x + y = 15$$

3. The "two equations type is worth **5** points". This tells us that every time we consider the points of the type **x** of questions, we will use **5x**. In other words, the total points in the test from questions with two equations are their number **x** multiplied by **5**.

4. "The three equations type is worth **10** points". This tells us that every time we consider the points of the type **y** of questions, we will use **10y**. In other words, the total points in the test from questions with three equations are their number **y** multiplied by **10**.

5. "Total points on test are **100**." This means that when we add all **5x** and **10y**, which we discussed above, they will add up to **100**. In an equation form this looks like:

$$5x + 10y = 100$$

Since the above facts are related, then we can use Systems of Equations to find out the number of each type of question. So, the following is our System of Equations:

$$\begin{cases} x + y = 15 \\ 5x + 10y = 100 \end{cases}$$

Let's use the **"Substitution Method"** to solve this system of equations. We will first isolate, let's say, **x** in the first equation:

$$\begin{cases} x = 15 - y \\ 5x + 10y = 100 \end{cases}$$

Now we will substitute **x** in the other equation, which is:

$$5x + 10y = 100$$

Let's substitute **x**:

$$5(15 - y) + 10y = 100$$

Let's solve for **y**:

$$75 - 5y + 10y = 100$$

$$-5y + 10y = 100 - 75$$

$$5y = 25$$

$$y = \frac{25}{5}$$

$$y = 5$$

So, we found the value of **y**. Let's substitute this in the equation where **x** is isolated:

$$x = 15 - y$$

$$x = 15 - 5$$

$$x = 10$$

We found that:

1. There are **10** questions about Systems of Equations with **2** equations (**x = 10**), and,

2. There are **5** questions about Systems of Equations with **3** equations (**y = 5**).

Check your work

Let's use **x = 10** and **y = 5** and confirm that our equations are in balance. Our System of Equations is:

$$\begin{cases} x + y = 15 \\ 5x + 10y = 100 \end{cases}$$

After we plug in the values for **x** and **y**:

$$\begin{cases} 10 + 5 = 15 \\ 5 \times 10 + 10 \times 5 = 100 \end{cases}$$

$$\begin{cases} 15 = 15 \\ 100 = 100 \end{cases}$$

This proves that our calculations were correct.

24. The two-digit number

Let's say that there is a two-digit number, and when you add its digits the sum is **10**. When you reverse the order of digits and subtract the original two-digit number, the difference is **72**.

Which number is it?

Solution

Let's understand and extract the facts that are given to us.

From the wording of our question, we understand that there are a couple of details given to us:

1. "there is a two-digit number,"
2. "the sum of its digits is **10**",
3. "reverse the order of digits", and
4. "When we subtract the **original** two-digit number from the **reversed** number the difference is **72**".

Now, let's "translate" the facts given in sentences into a mathematical notation:

1. Since we are looking for each digit of a two-digit number AND they are unknown to us at this moment, we will simply use two math symbols for unknowns such as **x** and **y**. Let's use:
   - **x** for the "tens" part of that number, and
   - **y** for the "ones" part of that number.

So, our two digit number looks like **xy**.

2. The fact "the sum of its digits is **10**", in Math notation can be written like:

$$x + y = 10$$

3. "Reverse the order of digits". This means that **x** and **y** will switch places and the new reversed number will look like **yx**.

4. When we subtract the original two-digit number from the reversed number the difference is **72**. This is the trickiest part to understand and requires more explanation.

It is clear that subtraction is involved, but what you **MUST NOT DO** is:

$$yx - xy = 72 \qquad \text{-- Wrong!!}$$

It is wrong because this will not help you find **x** and **y**, but can only help to check the result at the end. What you need is a way to separate the number **xy** into its constituents **x** and **y** so that you can isolate and find their values separately. Keep reading and it will be clear how and why.

Recall from first grade that the digits in a number have a place value. The two digit numbers have a "ten" and "ones", remember? We have to bear this in mind when we do the subtraction.

So, our **xy** number will be:

$$10x + y$$

And our reverse **yx** number will be:

$$10y + x$$

Now we can do the subtraction from fact number **4**:

$$(10y + x) - (10x + y) = 72$$

Let's simplify this equation by first removing the parenthesis:

$$10y + x - 10x - y = 72$$
$$9y - 9x = 72$$

If we divide every term by **9**, we get:

$$y - x = 8$$

Since the above facts are related, then we can use Systems of Equations to find the digits of our Two-Digit Number. So, the following is our System of Equations:

$$\begin{cases} x + y = 10 \\ y - x = 8 \end{cases}$$

Let's use the **"Addition Method"** to solve this system of equations. We add the sides:

$$(x + y) + (y - x) = 10 + 8$$

$$x + y + y - x = 18$$

$$2y = 18$$

$$y = \frac{18}{2}$$

$$y = 9$$

So, we found the value of **y**. Let's substitute this in, let's say, the first equation:

$$x + y = 10$$

$$x = 10 - y$$

$$x = 10 - 9$$

$$x = 1$$

So, we found **x** and **y** which represent the digits of our **xy** number, and our number is **19**.

Check your work

Let's use **x = 1** and **y = 9** and confirm that our equations are in balance.

Our System of Equations is:

$$\begin{cases} x + y = 10 \\ y - x = 8 \end{cases}$$

After we plug in the values for **x** and **y**:

$$\begin{cases} 1 + 9 = 10 \\ 9 - 1 = 8 \end{cases}$$

$$\begin{cases} 10 = 10 \\ 8 = 8 \end{cases}$$

This proves that our calculations while solving the Systems of Equations were correct. But how about the fact: "When we subtract the original two-digit number from the reversed number the difference is **72**!! Let's check this too:

If we found that **x = 1** and **y = 9** then our two-digit number is **19**. The reverse then is **91**. The difference will be:

$$91 - 19 = 72$$

This confirms that everything we did in this example is correct.

25. The three-digit number

Let's say that there is a three-digit number, such that when you add its digits the sum is **6**. If you reverse the order of digits then the value increases for **198**.

Which number is it?

Solution

Let's understand and extract the facts that are given to us.

From the wording of our question, we understand that there lots of details given to us:

1. "There is a three digit number,"
2. "The sum of its digits is **6**",
3. "Reverse the order of digits", and
4. "The value increases for **198**" when we reverse the digits.

Now, let's "translate" the facts given in sentences into a mathematical notation:

1. Since we are looking for each digit of a three digit number AND they are unknown to us at this moment, we will simply use three math symbols for unknowns such as **x**, **y** and **z**. Let's use:

   - **x** - for the **"hundreds"** part of that number
   - **y** - for the **"tens"** part of that number, and
   - **z** - for the **"ones"** part of that number.

So, our three digit number looks like **xyz**.

2. The fact that the sum of its digits is **6** , in math notation can be written like:

$$x + y + z = 6$$

3. The fact: "Reverse the order of digits", means that **x** and **z** will switch places and the new reversed number will look like **zyx**.

4. The fact: "the value increases for **198**". This part is a tricky part to understand and requires more explanation, so pay attention:

One thing you **MUST NOT DO** is:

$$zyx = xyz + 198 \quad \text{-- Wrong!!}$$

It is wrong because this will not help you find *x*, *y* and **z**, but can only help to check the result at the end. What you need is a way to separate the number **xyz** into its constituents **x**, **y** and **z**.

Recall from first grade that the digits in a number have a place value. The three digit numbers have a "**hundred**", a "**ten**" and "**ones**", remember? We must not forget this fact about the place value.

So, our **xyz** number will be:

$$100x + 10y + z$$

And our reverse **zyx** number will be:

$$100z + 10y + x$$

Now we can write the fact number **4** in math notation like:

$$100z + 10y + x = 100x + 10y + z + 198$$

Let's simplify this equation by first bringing the variables from the right side of equation to the left side:

$$100z + 10y + x - 100x - 10y - z = 198$$

After we add the like terms:

$$99z - 99x = 198$$

If we divide every term by **99**, we get:

$$z - x = 2$$

Surely you noticed that **y** got eliminated (**10y − 10y = 0**). The question is what does this mean for us? We will explain this in a little while.

Since the above facts are related, then we can use Systems of Equations to find out the digits of our three-digit number. So, the following is our System of Equations:

$$\begin{cases} x + y + z = 6 \\ z - x = 2 \end{cases}$$

Now let's explain the meaning of the fact that **y** got eliminated and the effect it has on our system of equations. It simply means that the second equation (condition) does not depend on **y**. In other Math words it means that **y** is independent variable here. In plain English, this means that **y** has more than one value and we can give arbitrary values to **y** from (**0…9**) and satisfy the condition: "If you reverse the order of digits then the value increases for **198**". Let's try some examples where **z** is bigger than **x** for **2** (because **z − x = 2**):

- ➢ For **y = 0** we get **301 = 103 + 198**
- ➢ For **y = 1** we get **311 = 113 + 198**
- ➢ For **y = 2** we get **321 = 123 + 198**
- ➢ For **y = 3** we get **331 = 133 + 198**
- ➢ For **y = 3** we get **634 = 436 + 198**

And there are more cases. Try some for yourself where **z − x = 2**.

However, we have the first condition or equation where the sum of its digits is **6**. As you can see from the examples above that satisfy the second condition, not all of them satisfy the first condition e.g. **436, 133, 113, 103** do not add to **6**.

So, let's solve our System of equation and see what happens:

$$\begin{cases} x + y + z = 6 \\ z - x = 2 \end{cases}$$

We will use the **"Substitution Method"** to solve this system of equations. Let's first isolate **z** in the second equation:

$$\begin{cases} x + y + z = 6 \\ z = 2 + x \end{cases}$$

Let's substitute **z** in the first equation:

$$x + y + z = 6$$
$$x + y + (2 + x) = 6$$
$$x + y + 2 + x = 6$$
$$2x + y + 2 = 6$$
$$2x + y = 6 - 2$$
$$2x + y = 4$$

Since **y** is independent variable, we will move it on the right side of equation and give arbitrary values to find **x**:

$$2x = 4 - y$$

Divide all terms by **2**:

$$x = 2 - \frac{1}{2}y$$

Let's find find now some numbers which satisfy both conditions: the sum is **6** and value increases by **198**:

Scenario 1:

For **y = 0** we get:

$$x = 2 - \frac{1}{2}y$$
$$= 2 - \frac{1}{2} \times 0$$
$$= 2 - 0$$
$$= 2$$

Let's find **z**:

$$z = 2 + x$$
$$= 2 + 2$$
$$= 4$$

So, we have **x = 2, y = 0, z = 4** and our number is **204**. This number satisfies both conditions:

$$2 + 0 + 4 = 6$$
$$402 = 204 + 198$$

Scenario 2:

For **y = 1** we get:

$$x = 2 - \frac{1}{2}y$$
$$= 2 - \frac{1}{2} \times 1$$
$$= 2 - \frac{1}{2}$$
$$= \frac{3}{2}$$

This is not a solution because **x** cannot be a fraction. It's a digit and must be a whole number.

Scenario 3:

For **y = 2** we get:

$$x = 2 - \frac{1}{2}y$$
$$= 2 - \frac{1}{2} \times 2$$
$$= 2 - 1$$
$$= 1$$

Let's find **z**:

$$z = 2 + x$$
$$= 2 + 1$$
$$= 3$$

So, we have **x = 1, y = 2, z = 3** and our number is **123**. This number satisfies both conditions:

$$1 + 2 + 3 = 6$$
$$321 = 123 + 198$$

Scenario 4:

For **y = 3** we get:

$$x = 2 - \frac{1}{2}y$$
$$= 2 - \frac{1}{2} \times 3$$
$$= 2 - \frac{3}{2}$$
$$= \frac{4 - 3}{2}$$
$$= \frac{1}{2}$$

This is not a solution because **x** cannot be a fraction. It's a digit and must be a whole number.

Scenario 5:

For **y = 4** we get:

$$x = 2 - \frac{1}{2}y$$
$$= 2 - \frac{1}{2} \times 4$$
$$= 2 - 2$$
$$= 0$$

Let's find **z**:

$$z = 2 + x$$
$$= 2 + 0$$
$$= 2$$

So, we have **x = 0, y = 4, z = 2** and our number is **042**. This number satisfies both conditions:

$$0 + 4 + 2 = 6$$
$$240 = 42 + 198$$

Scenario 6:

For **y = 5** we get:

$$x = 2 - \frac{1}{2}y$$

$$= 2 - \frac{1}{2} \times 5$$

$$= 2 - \frac{5}{2}$$

$$= -0.5$$

As you can see, for $y \in \{5, 6, 7, 8, 9\}$ we will get negative x, and that cannot be the solution for our system of equation.

We conclude that we have three solutions for our System of Equations, and they are:

1. $x = 2, y = 0, z = 4$ and our number is **204**.
2. $x = 1, y = 2, z = 3$ and our number is **123**.
3. $x = 0, y = 4, z = 2$ and our number is **042**.

26. The speed of Hudson River

Let's say that you decided to visit New York City, and in order to enjoy the Manhattan view, you decided to kayak on Hudson River. While kayaking downstream, towards downtown Manhattan, your average speed is $13\frac{mi}{hr}$. While kayaking upstream, towards uptown Manhattan, your average speed is $7\frac{mi}{hr}$.

What is the speed of Hudson River, and what would be your speed in still water?

Solution

Let's understand and extract the facts that are given to us.

We understand that there are many details given to us:

1. There is Hudson River speed and your kayaking speed. It is important to understand that if the water was still, then it would not make difference in your speed based on which direction you are kayaking. Since the Hudson River is flowing at certain speed in one direction, then it will either speed you up or slow you down in relation to, let's say, Manhattan buildings.
2. Downstream average speed is $13\frac{mi}{hr}$,
3. Upstream average speed is $7\frac{mi}{hr}$, and

Now, let's "translate" the facts given in sentences into a mathematical notation:

1. Since Hudson River speed and your kayaking speed are unknown to us at this moment, we will simply use two Math symbols for unknowns such as **x**, and **y**.
   Let's use:
     o **x** - for Hudson River speed, and
     o **y** - for your kayaking speed.
2. The fact "downstream average speed is $13\frac{mi}{hr}$", means that when you are kayaking in the direction of the stream, then your total speed in relation to the Manhattan

skyscrapers will be your kayak speed plus the Hudson River speed. In Math notation this fact can be written like:

$$y + x = 13$$

3. The fact "upstream average speed is $7\frac{mi}{hr}$", means that you are kayaking in the opposite direction of the water and this will make you slowdown in relation to the Manhattan skyscrapers. In math notation this fact can be written like:

$$y - x = 7$$

The negative sign $(-)$ of **x** indicates the opposite direction, and it will slow you down.

Since the above facts are related, then we can use Systems of Equations to find the speed of Hudson River, and your speed in still water. So, the following is our System of Equations:

$$\begin{cases} y + x = 13 \\ y - x = 7 \end{cases}$$

Let's use the **"Addition Method"** to solve this system of equations. We add the sides of equations:

$$(y + x) + (y - x) = 13 + 7$$
$$y + x + y - x = 20$$
$$2y = 20$$
$$y = \frac{20}{2}$$
$$y = 10$$

Let's substitute the value of **y** in the first equation:

$$y + x = 13$$
$$x = 13 - y$$
$$x = 13 - 10$$
$$x = 3$$

So, we found that speed of Hudson River is $3\frac{mi}{hr}$, and your kayaking speed in still water would be $10\frac{mi}{hr}$.

Check your work

Let's use **x = 3** and **y = 10** and confirm that our equations are in balance.

Our System of Equations is:

$$\begin{cases} y + x = 13 \\ y - x = 7 \end{cases}$$

After we plug in the values for **x** and **y**:

$$\begin{cases} 10 + 3 = 13 \\ 10 - 3 = 7 \end{cases}$$

$$\begin{cases} 13 = 13 \\ 7 = 7 \end{cases}$$

This proves that our calculations were correct.

27. The speed of a Mountain River

Let's say that you were kayaking on Mountain River and for **8 hrs** you covered **56 miles** while traveling downstream. It took you **14 hrs** to return to the starting position.

What would be your speed in still water, and what was the speed of the current?

Solution

Let's understand and extract the facts that are given to us.

We understand that there few details given to us:

1. "There is your kayaking speed and current speed (speed of a river)". It is important to understand that if the water was still, then it would not make difference in your speed based on which direction you are kayaking. Since the mountain river is flowing at certain speed in one direction, then it will either speed you up, if you travel downstream, or slow you down if you travel upstream.
2. "For **8 hrs** you covered **56 mi** while traveling downstream",
3. "**14 hrs** to return".

Now, let's "translate" the facts given in sentences into a mathematical notation:

1. Since the current and your speed are unknown to us at this moment, we will simply use two math symbols for unknowns such as **x**, and **y**.
   Let's use:
   - **x** - for your speed, and
   - **y** - for the current.
2. How can the fact: "for **8 hrs** you covered **56 miles** while traveling downstream", be written in Math notation? Since we have to find the speeds in miles-per-hour $\left(\frac{\text{mi}}{\text{hr}}\right)$, then we should write the downstream travel in the same units. So, if you traveled **56 mi** in **8 hrs**, then in one hour you traveled:

$$\frac{56 \text{ mi}}{8 \text{ hr}} = 7 \frac{\text{mi}}{\text{hr}}$$

This means that your speed plus the speed of the river resulted in $7 \frac{\text{mi}}{\text{hr}}$. In Math notation this fact can be written like:

$$x + y = 7$$

3. The fact "**14 hrs** to return" means that you have traveled the same **56 miles** but in the opposite direction, upstream. This will make you slow down. In miles-per-hour $\left(\frac{\text{mi}}{\text{hr}}\right)$, this would be:

$$\frac{56 \text{ mi}}{14 \text{ hr}} = 4 \frac{\text{mi}}{\text{hr}}$$

In math notation this fact can be written like:

$$x - y = 4$$

The negative sign $(-)$ of **y** indicates the opposite direction, and it will slow you down.

Since the above facts are related, then we can use Systems of Equations to find the speed of the Mountain River and your speed in still water. So, the following is our System of Equations:

$$\begin{cases} x + y = 7 \\ x - y = 4 \end{cases}$$

Let's use the **"Addition Method"** to solve this system of equations. We add the sides:

$$(x + y) + (x - y) = 7 + 4$$

$$x + y + x - y = 11$$

$$2x = 11$$

$$x = \frac{11}{2}$$

$$x = 5.5$$

Let's substitute the value of **x** in the first equation:

$$x + y = 7$$

$$y = 7 - x$$

$$y = 7 - 5.5$$

$$y = 1.5$$

So, we found that speed of the current is $\mathbf{1.5}\ \frac{\mathbf{mi}}{\mathbf{hr}}$, and your kayaking speed in still water would be $\mathbf{5.5}\ \frac{\mathbf{mi}}{\mathbf{hr}}$.

Check your work

Let's use $\mathbf{x = 5.5}$ and $\mathbf{y = 1.5}$ and confirm that our equations are in balance.

Our System of Equations is:

$$\begin{cases} x + y = 7 \\ x - y = 4 \end{cases}$$

After we plug in the values for **x** and **y**:

$$\begin{cases} 5.5 + 1.5 = 7 \\ 5.5 - 1.5 = 4 \end{cases}$$

$$\begin{cases} 7 = 7 \\ 4 = 4 \end{cases}$$

This proves that our calculations were correct.

28. Investing in stocks

Let's say that you and your best friend are discussing how to invest your savings in a stock market. After a small research, both of you decided to invest your money in Apple and Google stocks. You invested **$8,000** and bought **30** Apple shares and **40** Google shares. Your best friend invested **$5,700** and bought **20** Apple shares and **30** Google shares.

Given that both of you paid the same price for all of your shares, find out how much was the price of Apple shares and how much was the price of Google shares?

Solution

Let's understand and extract the facts that are given to us.

From the wording of our question, we understand that there are many details given to us:

1. "Two types of shares (stocks): Apple shares and Google shares,"
2. " **30** Apple shares and **40** Google shares total **$8,000**",
3. " **20** Apple shares and **30** Google shares total **$5,700**".

Now, let's "translate" the facts given in sentences into a mathematical notation:

1. Since we need to find out the prices of Apple and Google shares AND they are unknown to us at this moment, we will simply use two Math symbols for unknowns such as **x** and **y**.
   Let's use:
   - **x** - for Apple shares, and
   - **y** - for Google shares.
2. The fact that **30** Apple shares and **40** Google shares total **$8,000**, in math notation can be written like:

$$30x + 40y = 8000$$

3. The fact that **20** Apple shares and **30** Google shares total $\mathbf{\$5,700}$, in Math notation can be written like:

$$20x + 30y = 5700$$

Since the above facts are related, then we can use Systems of Equations to find out the prices of Apple and Google shares. So, the following is our System of Equations:

$$\begin{cases} 30x + 40y = 8000 \\ 20x + 30y = 5700 \end{cases}$$

Let's use the **"Addition Method"** to solve this system of equations. Recall from previous examples that first we need to transform one of the equations in a such a way that when we add the equations side by side, one of the variables gets eliminated and we solve for the remaining one. Let's say that we want to eliminate the variable **x**. In order to achieve that, we need to make the coefficients of **x** in both equations of the same value but with opposite signs. One way to do this is to multiply the second equation by $(\mathbf{-1.5})$:

$$\begin{cases} 30x + 40y = 8000 \\ 20x + 30y = 5700 \qquad /\times (-1.5) \end{cases}$$

Multiply all terms by $(\mathbf{-1.5})$:

$$\begin{cases} 30x + 40y = 8000 \\ (-1.5) \times 20x + (-1.5) \times 30y = (-1.5) \times 5700 \end{cases}$$

$$\begin{cases} 30x + 40y = 8000 \\ -30x - 45y = -8550 \end{cases}$$

Add equations side by side:

$$(30x + 40y) + (-30x - 45y) = 8000 + (-8550)$$

Let's remove parentheses:

$$30x + 40y - 30x - 45y = 8000 - 8550$$

Add the "like" terms`:

$$-5y = -550$$

$$y = \frac{-550}{-5}$$

$$y = 110$$

So, we found the value of **y**. Let's substitute this in any equation and find **x**. Let's choose the first equation:

$$30x + 40y = 8000$$

$$30x + 40 \times 110 = 8000$$

$$30x + 4400 = 8000$$

$$30x = 8000 - 4400$$

$$30x = 3600$$

$$x = \frac{3600}{30}$$

$$x = 120$$

We found that the price of one Apple share is **$120** and the price of one Google share is **$110**.

Check your work

Let's use **x = 120** and **y = 110** and confirm that our equations are in balance.

Our System of Equations is:

$$\begin{cases} 30x + 40y = 8000 \\ 20x + 30y = 5700 \end{cases}$$

After we plug in the values for **x** and **y**:

$$\begin{cases} 30 \times 120 + 40 \times 110 = 8000 \\ 20 \times 120 + 30 \times 110 = 5700 \end{cases}$$

$$\begin{cases} 8000 = 8000 \\ 5700 = 5700 \end{cases}$$

This proves that our calculations were correct.

29. College tuition

Let's say that a particular University decided to offer all their courses online as well, and students can earn their degree **100%** online. The tuition for the online degree is **20%** cheaper than the traditional on campus degree.

Last semester the college calculated that the revenue from the **70** Online Degree Students and **100** traditional students was **$156,000**. This semester the Online Degrees gained in popularity and the number of students seeking an online degree doubled and the total revenue increased to **$212,000**.

How much does the university charge per semester for an online degree, and how much for a traditional or on campus degree?

Solution

Let's understand and extract the facts that are given to us. From the wording of our question, we understand that there are many details given to us:

1. "Two types of college degrees: Online degree and Traditional degree"
2. The tuition of "**70** Online Degree Students and **100** traditional students totaled **$156,000**",
3. "students seeking an online degree doubled and the total revenue increased to **$212,000**",
4. "The tuition for the online degree is **20%** cheaper than the traditional"

Now, let's "translate" the facts given in sentences into a mathematical notation:

1. Since we need to find out the price per semester for an online degree and on campus degree AND they are unknown to us at this moment, we will simply use two math symbols for unknowns such as **x** and **y**. Let's use:
   - **x** - for Online College degree, and
   - **y** - for traditional, on campus degrees.

2. The fact that **70** Online Degree Students and **100** traditional students paid in total **$ 156, 000**, in math notation can be written like:

$$70x + 100y = 156{,}000$$

3. The fact "students seeking an online degree doubled and the total revenue increased to **$ 212, 000**" means that this semester there are **140** students seeking an online degree and, of course, revenue increased. Note that the number of on campus students remains the same. In math notation this fact can be written like:

$$140x + 100y = 212{,}000$$

4. The fact "The tuition for the online degree is **20%** cheaper than the traditional" does not help us to find the tuition prices, but it can help to confirm their relationship if you want to. Do not get confused by this. We are trying to determine the tuition prices based on the number of students and revenue, and not based on the relationship of the online degree tuition vs. on campus tuition. Hence, we can call this fact just "noise" and not use it while we are solving the problem.

Since the above facts are related, then we can use Systems of Equations to find out tuition prices for our two types of college degrees. So, the following is our System of Equations:

$$\begin{cases} 70x + 100y = 156{,}000 \\ 140x + 100y = 212{,}000 \end{cases}$$

Let's use the **"Addition Method"** to solve this system of equations. Recall from previous examples that first we need to transform one of the equations in a such a way that when we add the equations side by side, one of the variables gets eliminated and we solve for the remaining one. Let's say that we want first to eliminate **y**. In order to achieve that, we need to make the coefficients of **y** in both equations of the same value but with opposite signs. One way to do this is to multiply the first equation by (-1) and change the sign only of the coefficient of **y** variable, which is **100** in both equations:

$$\begin{cases} 70x + 100y = 156{,}000 \quad /\times (-1) \\ 140x + 100y = 212{,}000 \end{cases}$$

Multiplying all terms of the first equation by (-1) changes their sign:

$$\begin{cases} -70x - 100y = -156000 \\ 140x + 100y = 212000 \end{cases}$$

Add equations side by side:

$$(-70x - 100y) + (140x + 100y) = -156{,}000 + 212{,}000$$

Let's remove parenthesis:

$$-70x - 100y + 140x + 100y = 56{,}000$$

Add the "like" terms`:

$$70x = 56{,}000$$

$$x = \frac{56{,}000}{70}$$

$$x = 800$$

So, we found the value of **y**. Let's substitute this in any equation and find **x**. Let's choose the second equation:

$$140x + 100y = 212{,}000$$

$$140 \times 800 + 100y = 212{,}000$$

$$112{,}000 + 100y = 212{,}000$$

$$100y = 212{,}000 - 112{,}000$$

$$100y = 100{,}000$$

$$y = \frac{100{,}000}{100}$$

$$y = 1000$$

We found that the college charges $ **800** per semester for an online degree, and $ **1,000** per semester for an on campus degree.

Check your work

Let's use **x = 800** and **y = 1000** and confirm that our equations are in balance.

Our System of Equations is:

$$\begin{cases} 70x + 100y = 156{,}000 \\ 140x + 100y = 212{,}000 \end{cases}$$

After we plug in the values for **x** and **y**:

$$\begin{cases} 70 \times 800 + 100 \times 1000 = 156{,}000 \\ 140 \times 800 + 100 \times 1000 = 212{,}000 \end{cases}$$

$$\begin{cases} 70x + 100y = 156{,}000 \\ 140x + 100y = 212{,}000 \end{cases}$$

This proves that our calculations about solving the Systems of Equations were correct. But, how about the facts: "The tuition for the online degree is **20%** cheaper than the traditional". Is this still holding true? Let's check this too.

First you need to know the formula of how to determine how many percent **p** is some quantity (part **P**) of some other quantity (Total **T**):

$$p = \frac{P \times 100}{T}$$

In our case, the online tuition is **Part** of **Total** tuition (campus tuition), which means our **P = 800** and **T = 100** and **p =?**

Let's plug in these values in the formula.

$$p = \frac{800 \times 100}{1000} = \frac{80{,}000}{1000} = 80$$

Since online tuition is **p = 80**% of campus tuition, it means that it is for **20**% cheaper.

30. Diluting full fat milk

Let's say that you have milk with **12%** fat, and milk with **2%** fat. Let's say that you want to mix them so that you can get milk which will contain **6%** fat.

If you want to make **50 gal** of **6%** fat Milk, then how much of **12%** and **2%** fat milk you need to mix?

Solution

Let's understand and extract the facts that are given to us.

From the wording of our question we understand that there few of details given to us:

1. "Two types of milk: milk with **2%** fat and milk with **12%** fat,"
2. "You want to make **50 gal** of milk",
3. "You want to make milk with **6%** fat",

Now, let's "translate" the facts given in sentences into a mathematical notation:

1. Since we need to find out how much of **12%** fat milk and how much of **2%** fat milk we need to mix AND they are unknown to us at this moment, we will simply use two math symbols for unknowns such as **x** and **y** to mark them. Let's use:
   - **x** - for **2%** fat milk, and
   - **y** - for **12%** fat milk.
2. If "you want to make **50 gal** of milk", all you have to do is mix some **2%** and **12%** fat milk so that you get **50 gal** milk. In math notation this can be written like:

$$x + y = 50$$

3. "Milk with **6%** fat." This fact is a bit harder to understand and explain, so pay attention. You don't want only **6%** fat milk, but you want your **50 gal** milk to

have **6%** fat. Since we are mixing two types of milk in different quantities, we must put this **6%** fat in context or relation to the amounts of milk we are mixing and the total **50 gal** we are getting from the mixture.

We should ask the following question: **How many gallons of fat are in 50 gal milk, if this milk will contain 6% fat?**

Let's find this out:

$$50 \text{ gal} \times 6\% = 50 \times 0.06 = 3$$

So, there are **3 gal** fat in **50 gal** milk, if the milk contains **6%** fat. This is important to know because we are combining the fat of **2%** fat milk AND the fat of **12%** fat milk so that we can get **3 gal** of fat. In math notation this finding can be written like:

$$2\%x + 12\%y = 3 \text{ gal} \quad \text{of fat, or}$$

$$0.02x + 0.12y = 3$$

Since the above facts are related, then we can use Systems of Equations to find out how much of each type of milk we need to mix. So, the following is our System of Equations:

$$\begin{cases} x + y = 50 \\ 0.02x + 0.12y = 3 \end{cases}$$

Since working with decimal numbers is a little bit hard, let's remove them by multiplying the second equation by **50**:

$$\begin{cases} x + y = 50 \\ 0.02x + 0.12y = 3 \quad /\times 50 \end{cases}$$

After we multiply all terms by **50**:

$$\begin{cases} x + y = 50 \\ x + 6y = 150 \end{cases}$$

Let's use the **"Subtraction Method"** to solve this system of equations and subtract the first equation from the second equation:

$$(x + 6y) - (x + y) = 150 - 50$$

Let's remove parenthesis:

$$x + 6y - x - y = 100$$

Add the "like" terms`:

$$5y = 100$$

$$y = \frac{100}{5}$$

$$y = 20$$

So, we found the value of **y**. Let's substitute this in any equation and find **x**. Let's choose the first equation:

$$x + y = 50$$

$$x = 50 - y$$

$$x = 50 - 20$$

$$x = 30$$

We found that we need to mix **30 gal** of **2%** fat milk AND **20 gal** of **12%** fat milk so that at the end we have **50 gal** of milk with **6%** fat.

Check your work

Let's use **x = 30** and **y = 20** and confirm that our equations are in balance.

Our System of Equations is:

$$\begin{cases} x + y = 50 \\ 0.02x + 0.12y = 3 \end{cases}$$

After we plug in the values for **x** and **y**:

$$\begin{cases} 30 + 20 = 50 \\ 0.02 \times 30 + 0.12 \times 20 = 3 \end{cases}$$

$$\begin{cases} 50 = 50 \\ 3 = 3 \end{cases}$$

This proves that our calculations were correct.

31. Movie popularity

Let's say that the newest movie will start to be showing this weekend on your neighborhood movie theater. The movie will be available in **2D** and **3D**. The price for a **2D** movie ticket will be **$14** and for a **3D** movie ticket will be **$17**.

Over a **3** day period (Fri, Sat and Sun), worldwide, **10** million people watched the movie and theater boxes collected **$158,000,000**.

How many people watched the movie in **2D** and how many watched it in **3D**?

Solution

Let's understand and extract the facts that are given to us. From the wording of our question, we understand that there are lots of details given to us:

1. "Two types of movie watchers: **2D** movie watchers and **3D** movie watchers"
2. "Two types of movies: **2D** movie and **3D** movie",
3. "Two types of movie ticket prices: **2D** movie ticket price and **3D** movie ticket price",
4. "Over a **3** day period (Fri, Sat and Sun), worldwide",
5. "The price for a **2D** movie ticket is **$14**",
6. "The price for a **3D** movie ticket is **$17**",
7. "Theater boxes collected **$158,000,000**",
8. "**10,000,000** people watched the movie".

Now, let's "translate" the facts given in sentences into a mathematical notation:

1. Since we need to find out how many people watched the movie in **2D** and how many watched it in **3D** AND they are unknown to us at this moment, we will simply use two math symbols for unknowns such as **x** and **y** to mark them.
   Let's use:
      o **x** - for the number of people that watched the movie in **2D**, and
      o **y** - for the number of people that watched the movie in **3D**.

2. The fact "Two types of movies: **2D** movie and **3D** movie" should be ignored because, as you will see below, it is not going to help us to solve the problem. All we care is that a **2D** ticket price is different from a **3D** ticket price.

3. The fact "Two types of movie ticket prices" should be ignored because we are interested in the actual ticket prices and not the number of ticket types.

4. The fact "Over a **3** day period (Fri, Sat and Sun), worldwide" should be ignored because it is irrelevant for our purposes. It is "noise" in the question.

We are going to combine facts **5**, **6** and **7**. Since we know the total dollars that theater boxes collected and we know the prices for **2D** and **3D** movie ticket, then these facts can be combined and be written like:

$$14x + 17y = \$ \, 158{,}000{,}000$$

8. The fact "**10,000,000** people watched the movie" tells us the number of **2D** and **3D** movie watchers combined, and in math notation can be written like:

$$x + y = 10{,}000{,}000$$

Since the above facts are related, then we can use Systems of Equations to find out how many people watched the movie in **2D** and how many watched it in **3D**. So, the following is our System of Equations:

$$\begin{cases} 14x + 17y = 158{,}000{,}000 \\ \quad\; x + y = 10{,}000{,}000 \end{cases}$$

Let's use the **"Substitution Method"** to solve this system of equations:

Let's isolate **x** in the second equation:

$$\begin{cases} 14x + 17y = 158{,}000{,}000 \\ \quad\; x = 10{,}000{,}000 - y \end{cases}$$

Let's substitute the **x** in the first equation with the value of **x** in the second equation:

$$14 \times (10{,}000{,}000 - y) + 17y = 158{,}000{,}000$$

$$14 \times 10{,}000{,}000 - 14y + 17y = 158{,}000{,}000$$

$$140{,}000{,}000 + 3y = 158{,}000{,}000$$

$$3y = 158{,}000{,}000 - 140{,}000{,}000$$

$$3y = 18{,}000{,}000$$

$$y = \frac{18{,}000{,}000}{3}$$

$$y = 6{,}000{,}000$$

So, we found the value of **y**. We can substitute this in any equation and find **x**. Let's choose the equation where we isolated **x**:

$$x = 10{,}000{,}000 - y$$

$$x = 10{,}000{,}000 - 6{,}000{,}000$$

$$x = 4{,}000{,}000$$

We found that **4** million people watched the movie in **2D** and **6** million people watched it in **3D**.

Check your work

Let's use $x = \mathbf{4{,}000{,}000}$ and $y = \mathbf{6{,}000{,}000}$ and confirm that our equations are in balance.

Our System of Equations is:

$$\begin{cases} 14x + 17y = 158{,}000{,}000 \\ x + y = 10{,}000{,}000 \end{cases}$$

After we plug in the values for **x** and **y**:

$$\begin{cases} 14 \times 4{,}000{,}000 + 17 \times 6{,}000{,}000 = 158{,}000{,}000 \\ 4{,}000{,}000 + 6{,}000{,}000 = 10{,}000{,}000 \end{cases}$$

$$\begin{cases} 158{,}000{,}000 = 158{,}000{,}000 \\ 10{,}000{,}000 = 10{,}000{,}000 \end{cases}$$

This proves that our calculations were correct.

32. Internet speed

Let's say that you have a very fast Internet connection. You start to download a zip file that contains **7** songs and **2** video clips and it takes **19 sec** to fully download them. Another zip file that you downloaded had **9** songs and **4** video clips and it took **33 sec** to download.

If the songs were of equal size, and videos were of equal size, then how long would it take to download a zip file with only **1** song and **1** video clip?

Solution

As always, first we must understand and extract the facts that are given to us.

From the wording of our question, we understand that there are few details given to us:

1. "Two types of content: songs and videos."
2. " **7** songs and **2** video clips it takes **19 sec**" to download,
3. " **9** songs and **4** video clips and it took **33 sec**" to download,
4. "How long would it take to download **1** song and **1** video"?

Now, let's "translate" the facts given in sentences into a mathematical notation:

1. Since we need to find out how long would it take to download **1** song and **1** video AND they are unknown to us at this moment, we will simply use two math symbols for unknowns such as **x** and **y** to mark them.
   Let's use:
   - **x** - for the time it takes to download a song, and
   - **y** - for the time it takes to download a video.
2. The fact " **7** songs and **2** video clips it takes **19 sec**" in math notation can be written like:

$$7x + 2y = 19$$

3. The fact " **9** songs and **4** video clips took **33 sec**" in Math notation can be written like:

$$9x + 4y = 33$$

4. Our question is: "How long would it take to download **1** song and **1** video?" Note that in order to find out how long it takes to download a song and a video TOGETHER, we need to find out first how long it takes to download them SEPARATELY. In other words, we need to find **x** and **y**, and then calculate **x + y**.

Since the above facts are related, then we can use Systems of Equations to find out how long would it take to download **1** song and **1** video SEPARATELY. So, the following is our System of Equations:

$$\begin{cases} 7x + 2y = 19 \\ 9x + 4y = 33 \end{cases}$$

Let's use the **"Addition Method"** to solve this system of equations:

Let's multiply the first equation by (-2) (by now, you should now the reason why we do this. If not go back to previous examples):

$$\begin{cases} 7x + 2y = 19 \quad /\times (-2) \\ 9x + 4y = 33 \end{cases}$$

$$\begin{cases} -14x - 4y = -38 \\ 9x + 4y = 33 \end{cases}$$

Let's add the equations side by side:

$$(-14x - 4y) + (9x + 4y) = -38 + 33$$

Let's remove the parenthesis, and then add the "like" terms:

$$-14x - 4y + 9x + 4y = -38 + 33$$

$$-5x = -5$$

$$x = \frac{-5}{-5}$$

$$x = 1$$

Let's substitute **x** in any equation with the value of **1** so that we can find the value of **y**. Let's pick the second equation:

$$9x + 4y = 33$$

$$9 \times 1 + 4y = 33$$

$$9 + 4y = 33$$

$$4y = 33 - 9$$

$$4y = 24$$

$$y = \frac{24}{4}$$

$$y = 6$$

We found that it takes **1** second to download a song and **6** seconds to download a video clip. Remember, this is when the zip file contains ONLY **1** song or **1** video clip. If the song and the video are packaged into **1** zip file, then the download time will be:

$$x + y = 1 + 6 = 7 \text{ sec}$$

So, to answer our question: It will take **7** seconds to download the zip file with **1** song and **1** video clip.

Check your work

Let's use **x = 1** and **y = 6** and confirm that our equations are in balance.

Our System of Equations is:

$$\begin{cases} 7x + 2y = 19 \\ 9x + 4y = 33 \end{cases}$$

After we plug in the values for **x** and **y**:

$$\begin{cases} 7 \times 1 + 2 \times 6 = 19 \\ 9 \times 1 + 4 \times 6 = 33 \end{cases}$$

$$\begin{cases} 19 = 19 \\ 33 = 33 \end{cases}$$

This proves that our calculations were correct.

33. Deciding which Car to buy

Let's say that you decided to buy a new car, but you are debating whether you should get a hybrid car, which is more expensive but uses less gas, or buy a regular car, which is cheaper but uses more gas for the same miles driven.

The new hybrid car costs $\$27,000$, and if you buy this one, you will spend a total of $\$800$ on gas every year.

The regular car costs $\$22,500$, but if you buy this one, you will spend a total of $\$1,200$ on gas every year.

It is clear that if you buy a hybrid car you will pay a little more up front but you will save on gas year-after-year, and it can turn out to be a cheaper choice based on the miles you drive every year.

If you drive the same miles every year, and the price of gas does not change, then after how many years the total cost (car + gas) of the two cars will be the same? And how much would that total cost be?

Solution

As always, first we must understand and extract the facts that are given to us.

From the wording of our question, we understand that there many details given to us:

1. "Two types of cars: a Regular Car and a Hybrid Car."
2. The question is: **"how many years"** and **"how much would be the total cost"**,
3. "You drive the same miles every year, and the price of gas does not change",
4. "The hybrid car costs $\$27,000$", and "you will spend $\$800$ on gas every year"
5. "The regular car costs $\$22,500$" and "you will spend $\$1,200$ on gas every year"

Now, let's "translate" the facts given in sentences into a mathematical notation:

1. The fact that we have "two types of cars: a Regular Car and a Hybrid Car" is good to bear in mind, but cannot help to answer our questions listed in fact number **2**. Hence, we will not try to write anything in math notation about this fact.

2. Since we need to find out how many years and the cost AND they are unknown to us at this moment, we will simply use two math symbols for unknowns such as **x** and **y** to mark them.

   Let's use:
   - **x** - for the number of years it will take for the cost of the two cars to be the same, and
   - **y** - for the dollar amount of the cost, when the cost of the two cars becomes equal.

3. The fact that "you drive the same miles every year, and the price of gas does not change" is very important condition because if you drive different number of miles every year and gas price changes, then there is no way for us to answer the questions. We don't know how they will change after we buy the car. Hence, we will not try to write anything in math notation about this fact, but only understand that the assumption is that gas price and miles driven remain constant over the years.

4. Before we write the fact **4** in math notation, let's explain it in plain English what does it mean and how is it going to help us answer our questions. First, you have to buy the car and pay **$27,000**. As you use the car, you will need to buy gas. The way you use the car, you will spend **$800** every year. At the end of the **first** year you will have spent:

$$\$27000 + 1 \times \$800$$

At the end of the **second** year you will have spent:

$$\$27000 + 2 \times \$800$$

At the end of the **x** year you will have spent:

$$\$27000 + x \times \$800$$

Since we don't know how much that cost would be after **x** years, we will simply write **y**:

$$\$27000 + x \times \$800 = y$$

5. The same applies for the regular car as well. First, you have to buy the car and pay **$22,500**. As you use the car, you will need to buy gas. The way you use the car, you will spend **$1,200** every year. At the end of the **first** year you will have spent:

$$\$22500 + 1 \times \$1200$$

At the end of the **second** year you will have spent:

$$\$22500 + 2 \times \$1200$$

At the end of the **x** year you will have spent:

$$\$22500 + x \times \$1200$$

Since we don't know how much that cost would be after **x** years, we will simply write **y**:

$$\$22500 + x \times \$1200 = y$$

Pay attention, in both equations above we used the same symbol **y** for the total cost. At different years the cost of two cars will be different, **but**, and this is very important, at year **x** the cost will be the same and this is why we can use the same symbol **y** for both cars.

Since the above facts are related, then we can use Systems of Equations to find out **"how many years"** and **"how much would be the total cost"**. So, the following is our System of Equations:

$$\begin{cases} \$22,500 + \$1,200x = y \\ \$27,000 + \$800x = y \end{cases}$$

Let's fix it a bit by bringing variable **y** on the left side of equations, and plain numbers on the right side of equations:

$$\begin{cases} 1,200x - y = -22,500 \\ 800x - y = -27,000 \end{cases}$$

Let's use the **"Addition Method"** to solve this system of equations:

Let's multiply the first equation by (-1) (by now, you should now the reason why we do this. If not go back to previous examples):

$$\begin{cases} 1{,}200x - y = -22{,}500 \quad\quad /\times (-1) \\ 800x - y = -27{,}000 \end{cases}$$

$$\begin{cases} -1{,}200x + y = 22{,}500 \\ 800x - y = -27{,}000 \end{cases}$$

Let's add the equations side by side:

$$(-1200x + y) + (800x - y) = 22500 - 27000$$

Let's remove the parenthesis, and then add the "like" terms:

$$-1200x + y + 800x - y = 22500 - 27000$$

$$-400x = -4500$$

$$x = \frac{-4500}{-400}$$

$$x = 11.25$$

Let's substitute **x** in any equation with the value of **11.25** so that we can find **y**. Let's pick the first equation:

$$22500 + 1200x = y$$

$$22500 + 1200 \times 11.25 = y$$

$$22500 + 13500 = y$$

$$y = 36000$$

We found that it will takes **11.25** years, or **11** years and **3** months for the Total Cost of Ownership (TCO), based on car's purchase price and gas usage, to be the same. The total cost after **11.25** years for both cars would be $ **36,000**.

Check your work

Let's use **x = 11.25** and **y = 36,000** and confirm that our equations are in balance.

Our System of Equations is:

$$\begin{cases} \$22{,}500 + \$1{,}200x = y \\ \$27{,}000 + \$800x = y \end{cases}$$

After we plug in the values for **x** and **y**:

$$\begin{cases} \$22{,}500 + \$1{,}200 \times 11.25 = 36{,}000 \\ \$27{,}000 + \$800 \times 11.25 = 36{,}000 \end{cases}$$

$$\begin{cases} 36{,}000 = 36{,}000 \\ 36{,}000 = 36{,}000 \end{cases}$$

This proves that our calculations were correct.

34. Should you get a Master's Degree?

Let's say that there is a job out there that you would like to have one day. The job pays a salary of **$60,000** per year for candidates who have a Bachelor's Degree. If a candidate has a Master's degree, then the salary would be **20%** higher.

Imagine that you are about to finish your College and are debating whether to pursue a Master's Degree for the next two years or just get the job and start earning money.

It is clear that after two years, having completed your Master's, you will be getting higher Salary, but you will spend **$30,000** for your Master's Degree, and during the two years you won't earn a dime. On the other side, if you just get the job right after College, you will not spend **$30,000** for Master's Degree, you will start earning money right away, but you will earn less than if you had Master's Degree.

Does it make sense financially to pursue a Master's Degree?

Intuition tells us that if you have Master's Degree and earn higher Salary, even after you spend some money for your degree and don't work for two years, eventually you will be better off financially.

How many years will it take for you to make up the money, while getting "Master's Degree Salary", and break even (reaching the amount of money you would have earned right from out of College)? How much you would have earned in total up to that point?

Solution

As always, first we must analyze and understand the problem in depth: If you get the job having only Bachelor's Degree, you will start to earn money right away. If you pursue Master's degree, you will start earning money after two years. By this time you would have earned **$120,000** if you did not go for Master's. In addition, you will spend **$30,000** for your Master's Degree. It is true that after you finish your Master's you will earn more money, and eventually you might catch up or surpass that amount you would make with

Bachelor's only. But, when will you reach that point and how much money would that be? This is what we are trying to find out in this example.

Let us now extract the facts that are given to us:

From the wording of our question, we understand that there lots of details given to us:

1. "Two types of degrees: a bachelor's degree and master's degree."
2. Two types of salaries: **$60,000** and the one which is **20%** higher.
3. The question is: **"how many years will it take to make up the money?"**, and **"how much you would have earned in total up to that point?"**
4. "you will spend **$30,000** for your master degree",
5. "Two years to complete it and not be able to work".

Now, let's "translate" the facts given in sentences into a mathematical notation:

1. The fact that we have "Two types of degrees" is good to bear in mind, but cannot help to answer our questions listed in our third fact. Hence, we will not try to write anything in math notation about this fact. We are interested in something else and not in types of degrees.
2. The fact that we have "Two types of salaries" is important for us because we must use them in our calculation and answer our questions. The salary of a candidate with Bachelor's Degree is given in absolute terms, which is **$60,000**. The salary of a candidate with Master's Degree is given in relative terms, which is **20%** higher. So, first we need to find out how much is this salary in dollars:

When we say **20%** higher, we should right away ask: **20%** of what? Well, in our case the answer is **20%** of **$60,000**. Once we find how much this is, then we add this amount to the base salary of **$60,000** and we get the salary of a candidate with Master's Degree. In plain English, we can say: **$60,000** plus **20%** of **$60,000**. In Math we write this like:

$$\$60,000 + 20\% \times \$60,000 =$$

$$= \$60,000 + \frac{20}{100} \times \$60,000$$

$$= \$60,000 + 0.2 \times \$60,000$$

$$= \$60,000 + \$12,000$$

$$= \$72,000$$

Another way to calculate this is:

$$60,000 \times 1.2 = 72,000$$

So, the salary of a candidate with master's degree in absolute dollars is **$72,000** per year.

3. Since we need to find out **how many years** and the **total earnings** AND they are unknown to us at this moment, we will simply use two math symbols for unknowns such as **x** and **y** to mark them.
    Let's use:
    o   **x** - for the number of years it will take to make up the money and break even, and
    o   **y** - for the total dollar amount, when the two earners (the one with Bachelor's and the one with Master's) have reached the same amount of earnings.

In order to answer our questions, we need to understand how each of the two types of degrees will earn money over time.

Let's start with a case when you have only a Bachelor's Degree. At the end of the **first** year you would have earned:

$$\$60,000 \times 1 = \$60,000$$

At the end of the **second** year you would have earned:

$$\$60,000 \times 2 = \$120,000$$

At the end of the **x** year you would have earned:

$$\$60,000 \times x = y$$

Since we don't know how much you would have earned after **x** years, we simply wrote **y**.

Let's analyze the case when you have a master's degree. At the end of the **first** year you would have earned:

$$\$72,000 \times 1 = \$72,000$$

However, since we would like to compare with the case when you don't have master's degree, you have to take into account the facts number **4** and **5** above. You did spend $\$30,000$ and you did not work two years and so did not make $2 \times \$60,000 = \$120,000$. So, at the end of first year your financial situation will look like:

$$\$72,000 \times 1 - \$30,000 - \$120,000 =$$

$$= \$72,000 \times 1 - \$150,000$$

$$= -\$78,000$$

At the end of the **second** year your financial situation will look like:

$$\$72000 \times 2 - \$150,000 = -\$6,000$$

At the end of the **x** year your financial situation will look like:

$$\$72,000 \times x - \$150,000 = y$$

Pay attention, in both equations above we used the same symbol **y** for the total earnings. At different years the earnings of two cases (with/without Master's) will be different, **but**, and this is very important, at some year **x** the earnings will be the same and this is why we can use the same symbol **y** for both cases.

Since the above facts are related, then we can use Systems of Equations to find out **"how many years will it take to make up the money?"**, and **"how much you would have earned in total up to that point?"**

So, the following is our System of Equations:

$$\begin{cases} \$60,000\,x = y \\ \$72,000\,x - \$150,000 = y \end{cases}$$

Let's fix it a bit by bringing variable **y** on the left side of equations, and plain numbers on the right side of equations:

$$\begin{cases} 60{,}000\,x - y = 0 \\ 72{,}000\,x - y = 150{,}000 \end{cases}$$

Let's use the **"Addition Method"** to solve this system of equations:

Let's multiply the first equation by (-1) :

$$\begin{cases} 60{,}000\,x - y = 0 \qquad /\times (-1) \\ 72{,}000\,x - y = 150{,}000 \end{cases}$$

$$\begin{cases} -60{,}000\,x + y = 0 \\ 72{,}000\,x - y = 150{,}000 \end{cases}$$

Let's add the equations side by side:

$$(-60{,}000\,x + y) + (72{,}000\,x - y) = 0 + 150000$$

Let's remove the parenthesis, and then add the "like" terms:

$$-60{,}000\,x + y + 72{,}000\,x - y = 150{,}000$$

$$12{,}000x = 150{,}000$$

$$x = \frac{150{,}000}{12{,}000}$$

$$x = 12.5$$

Let's substitute **x** in any equation with the value of **12.5** so that we can find **y**. Let's pick the first equation:

$$60{,}000\,x = y$$

$$60{,}000 \times 12.5 = y$$

$$y = 750{,}000$$

We found that it will take **12.5** years for the earnings of the two cases (with/without Master's Degree) to be the same. The dollar amount of earnings at the end of **12** years and **6** months for both cases would be $ **750,000**.

So, after **12.5** years you will start reaping the financial benefits of your Master's Degree and be in a better financial standing than if you only had Bachelor's Degree.

Does it make sense financially to pursue a Master's Degree? Of course it does.

Check your work

Let's use **x = 12.5** and **y = 750,000** and confirm that our equations are in balance.

Our System of Equations is:

$$\begin{cases} \$60{,}000x = y \\ \$72{,}000x - \$150{,}000 = y \end{cases}$$

After we plug in the values for **x** and **y**:

$$\begin{cases} \$60{,}000 \times 12.5 = \$750{,}000 \\ \$72{,}000 \times 12.5 - \$150{,}000 = \$750{,}000 \end{cases}$$

$$\begin{cases} \$750{,}000 = \$750{,}000 \\ \$750{,}000 = \$750{,}000 \end{cases}$$

This proves that our calculations were correct.

35. Watch your Calorie intake

Many Fast Food restaurants offer meals where the same Sandwich is offered in combination with a small, or medium, or a large drink and Fries.

Let's say that your favorite Fast Food Restaurant offers:

1. **Kids meal:** Sandwich with small **3 oz** French fries and **16 oz** small Soda drink, which has in total **1072 cal**,

2. **Regular meal:** Sandwich with medium **4 oz** French fries and **20 oz** medium Soda drink, which has in total **1210 cal**, and

3. **Jumbo meal: 2** Sandwiches with large **6 oz** French fries and **28 oz** large Soda drink, which has in total **2096 cal**.

Since you already know that Fries and Soda are not healthy and add unnecessary calories but no nutrients, you decide to get only a Sandwich and a bottle of Water.

How many calories are in your lunch (Sandwich and Water)?

Solution

As usual, the very first thing we must do is to fully understand the problem and then extract the relevant facts that are available to us.

From the wording of our question, we understand that there are three types of menus (meals). Ultimately we want to know how many calories are in a Sandwich, and we are supposed to find this out based on the details given to us. Let's extract the facts:

1. All three meals consist of the same items: Sandwich, Fries and Soda drink. Only the amount per **oz** changes regarding the Fries and the Soda drink, and the number of Sandwiches per meal.

2. "Kids meal: Sandwich with small **3 oz** French Fries and **16 oz** small Soda drink, totaling **1072 cal**"

3. "Regular meal: Sandwich with medium **4 oz** French Fries and **20 oz** medium Soda drink, totaling **1210 cal**"

4. "Jumbo meal: **2** Sandwiches with medium **6 oz** French Fries and **28 oz** medium Soda drink, totaling **2096 cal**".

The question is to find out the calories of your lunch consisting of Sandwich and Water only. Since water has **0** calories, then we need to find only the calories of the Sandwich. In order to do this, we will need to find the calories of French Fries and Soda drink as well.

Now, let's "translate" the facts given in sentences into a mathematical notation:

1. Since we need to find out how many calories are in each item AND they are unknown to us at this moment, we will simply use three math symbols for unknowns such as **x**, **y** and **z** to mark them.
   Let's use:
      o **x** - for the calories in a Sandwich,
      o **y** - for the calories per **oz** in French Fries, and
      o **z** - for the calories per **oz** in Soda drink.

2. The fact "Sandwich with small **3 oz** French Fries and **16 oz** small Soda drink, totaling **1072 cal**" in math notation can be written as:

$$x + 3y + 16z = 1072$$

3. The fact "Sandwich with medium **4 oz** French Fries and **20 oz** medium Soda drink, totaling **1210 cal**" in math notation can be written as:

$$x + 4y + 20z = 1210$$

4. The fact " **2** Sandwiches with medium **6 oz** French Fries and **28 oz** medium Soda drink, totaling **2096 cal**" in math notation can be written as:

$$2x + 6y + 28z = 2096$$

Since the above facts are related, then we can use Systems of Equations to find out **"how many calories"** are in each item. So, the following is our System of Equations:

$$\begin{cases} x + 3y + 16z = 1072 \\ x + 4y + 20z = 1210 \\ 2x + 6y + 28z = 2096 \end{cases}$$

Let's use the **"Substitution Method"** to solve this system of equations:

We will isolate a variable in one equation, and substitute its value in another equation, until we can find a numeric value of one variable. Then we use this value to find the other values. Let us isolate **x** in the first equation:

$$\begin{cases} x = 1072 - 3y - 16z \\ x + 4y + 20z = 1210 \\ 2x + 6y + 28z = 2096 \end{cases}$$

Let's substitute the **x** in the second equation, which is:

$$x + 4y + 20z = 1210$$

After substitution:

$$(1072 - 3y - 16z) + 4y + 20z = 1210$$
$$1072 - 3y - 16z + 4y + 20z = 1210$$

Add the "like" terms:

$$y + 4z = 1210 - 1072$$
$$y + 4z = 138$$
$$y = 138 - 4z$$

Let's now substitute **x** in the third equation, which is:

$$2x + 6y + 28z = 2096$$
$$2 \times (1072 - 3y - 16z) + 6y + 28z = 2096$$
$$2144 - 6y - 32z + 6y + 28z = 2096$$

Add the "like" terms:

$$2144 - 4z = 2096$$

$$-4z = 2096 - 2144$$

$$-4z = -48$$

$$z = \frac{-48}{-4}$$

$$z = 12$$

We found that there are **12** calories per **oz** in Soda drink. Let's use this value and find **y**:

Let's use the value of **z** to find **y** in one of the transformed equations above. We found that:

$$y = 138 - 4z$$

Since **z = 12**, then:

$$y = 138 - 4 \times 12$$

$$y = 138 - 48$$

$$y = 90$$

We found that there are **90** calories per **oz** in French Fries. Now that we found the values of **y** and **z**, we can use them to find the value of **x**. We said that:

$$x = 1072 - 3y - 16z$$

After we substitute the values of **y** and **z**:

$$x = 1072 - 3 \times 90 - 16 \times 12$$

$$x = 1072 - 270 - 192$$

$$x = 610$$

We found that there are **610** calories in your Sandwich from your favorite Fast Food Restaurant! Since we said water has **0** calories, then your lunch has only the Sandwich calories of **610.**

Check your work

Let's use $x = 610, y = 90$ and $z = 12$ and confirm that our equations are in balance. Our System of Equations is:

$$\begin{cases} x + 3y + 16z = 1072 \\ x + 4y + 20z = 1210 \\ 2x + 6y + 28z = 2096 \end{cases}$$

After we plug in the values for **x** and **y**:

$$\begin{cases} 610 + 3 \times 90 + 16 \times 12 = 1072 \\ 610 + 4 \times 90 + 20 \times 12 = 1210 \\ 2 \times 610 + 6 \times 90 + 28 \times 12 = 2096 \end{cases}$$

$$\begin{cases} 1072 = 1072 \\ 1210 = 1210 \\ 2096 = 2096 \end{cases}$$

This proves that our calculations were correct.

36. Employee earnings

In one newly opened fast food Restaurant currently are working **6** people: **2** Cashiers, **3** Sandwich makers and **1** Manager. The Restaurant Payroll totals **$60** for **1 hr**.

Because they make yummy Sandwiches, their business is growing and they are hiring **1** more Cashier and **2** more Sandwich makers. The new payroll for **1 hr** will be **$86**.

The manager of the Restaurant is earning per hour exactly what one Cashier and one Sandwich maker are earning combined.

Find the dollar amount that each type of employee is paid per hour.

Solution

As usual, the very first thing we must do is to fully understand the problem and then extract the relevant facts that are available to us.

From the wording of our question, it is clear that:

1. There are three types of employees: Cashier, Sandwich maker and Manager.
2. Since there are three types of employees in the Restaurant, this means that there are three types of wages per hour, and this is what we need to find out.
3. "The payroll for **1 hr** totals **$60**." In other words, it costs the Restaurant **$60** per hour to have **2** Cashiers, **3** Sandwich makers and **1** Manager.
4. "The new payroll for **1 hr** will total **$86**." In other words, after they hire **1** more Cashier and **2** more Sandwich makers it will cost the Restaurant **$86** per hour to have **3** Cashiers, **5** Sandwich makers and **1** Manager.
5. "The manager is earning per hour exactly what a Cashier and sandwich maker are earning combined".

Now, let's "translate" the facts given in sentences into a mathematical notation:

1. Since we are looking for the wage per hour that each employee is earning, we will NOT notate the type of the employee but will notate the wages they earn in point **2** below.

2. Since we need to find out the wage per hour of each type of employee AND they are unknown to us at this moment, we will simply use three Math symbols for unknowns such as **c**, **s** and **m** to mark them. Let's use:

    c - For the wage per hour that the Cashier earns,

    s - For the wage per hour that the Sandwich Maker earns, and

    m - For the wage per hour that the Manager earns.

3. The fact "The payroll for **1 hr** totals **$60** for **2** Cashiers, **3** Sandwich makers and **1** Manager" in math notation can be written as:

$$2c + 3s + m = \$60$$

4. The fact "The payroll for **1 hr** totals **$86** for **3** Cashiers, **5** Sandwich makers and **1** Manager" in math notation can be written as:

$$3c + 5s + m = \$86$$

5. The fact "The manager is earning per hour exactly what a Cashier and Sandwich maker are earning combined" in Math notation can be written as:

$$c + s = m$$

Since the above facts are related, then we can use Systems of Equations to find out **"the dollar amount that each type of employee is paid per hour"**. So, the following is our System of Equations:

$$\begin{cases} 2c + 3s + m = 60 \\ 3c + 5s + m = 86 \\ c + s = m \end{cases}$$

Let's use the **"Substitution Method"** to solve this system of equations:

We need to isolate a variable in one equation, and substitute its value in another equation, until we can find a numeric value of one variable. Then we use this to find the other values.

As you can see, the variable **m** in the third equation is already isolated. So, we can use this and substitute it in both: first and second equation:

$$\begin{cases} 2c + 3s + (c + s) = 60 \\ 3c + 5s + (c + s) = 86 \end{cases}$$

Let's remove the parenthesis and add the like terms:

$$\begin{cases} 3c + 4s = 60 \\ 4c + 6s = 86 \end{cases}$$

Let's find the values for **c** and **s** in this system of equations and then we will use them to find the value of **m**.

Let's use the **"Addition Method"** to solve this System of Equations. Let's multiply the first equation by (-4) and the second equation by **3** so that we can eliminate variable **c**:

$$\begin{cases} 3c + 4s = 60 \quad /\times (-4) \\ 4c + 6s = 86 \quad /\times 3 \end{cases}$$

$$\begin{cases} -12c - 16s = -240 \\ 12c + 18s = 258 \end{cases}$$

Let's add the equations side by side:
$$(-12c - 16s) + (12c + 18s) = -240 + 258$$

Let's remove the parenthesis and add the like terms:
$$-12c - 16s + 12c + 18s = -240 + 258$$

$$2s = 18$$

$$s = \frac{18}{2}$$

$$s = 9$$

Let's pick the first equation and substitute $s = 9$ to find c:

$$3c + 4s = 60$$

$$3c + 4 \times 9 = 60$$

$$3c + 36 = 60$$

$$3c = 60 - 36$$

$$3c = 24$$

$$c = \frac{24}{3}$$

$$c = 8$$

Let's use $s = 9$ and $c = 8$ to find m:

$$c + s = m$$

$$8 + 9 = m$$

$$m = 17$$

So, we found that Cashier is being paid $\$8$ per hour, the Sandwich maker $\$9$ and the Manager $\$17$ per hour.

Check your work

Let's use $c = 8$, $s = 9$ and $m = 1$ and confirm that our equations are in balance.

Our System of Equations is:

$$\begin{cases} 2c + 3s + m = 60 \\ 3c + 5s + m = 86 \\ c + s = m \end{cases}$$

After we plug in the values for **x** and **y**:

$$\begin{cases} 2 \times 8 + 3 \times 9 + 17 = 60 \\ 3 \times 8 + 5 \times 9 + 17 = 86 \\ \qquad\quad 8 + 9 = 17 \end{cases}$$

$$\begin{cases} 60 = 60 \\ 86 = 86 \\ 17 = 17 \end{cases}$$

This proves that our calculations were correct.

37. Employee and Manager Raises

In a department store, the Manager and the **5** Cashiers together earn **$125** per hour.

Because they were running the business successfully, each Cashier will get a **5%** raise, and the Manager's raise will be the total of the **5** Cashier raises. Together they will earn **$135** per hour.

Find the dollar amount that the Manager and Cashiers are currently being paid per hour.

Solution

Let us first dissect the question so that we can fully understand the problem and then find the answers. Let's extract the relevant facts that are available to us.

From the wording of our question, it is clear that:

1. There are two types of employees: a Cashier and a Manager.
2. There are two types of wages per hour: Manager's wage per hour and Cashier's wage per hour. This is what we need to find out.
3. "The Manager and the **5** Cashiers together earn **$125** per hour."
4. "Each Cashier will get a **5%** raise."
5. "The Manager's raise will be the total of the **5** Cashier raises."
6. "Together they will earn **$135** per hour."

Now, let's "translate" the facts given in sentences into a mathematical notation:

1. The fact "Two types of employees: Cashier and Manager" is a fact that helps us understand the fact number **2**. Hence, we will not notate anything about this fact.
2. Since we need to find out the wage per hour of each type of employee AND they are unknown to us at this moment, we will simply use two math symbols for unknowns such as c and m to mark them. Let's use:

    c - For the wage per hour that the Cashier earns, and

m - For the wage per hour that the Manager earns.

3. The fact: "The Manager and the **5** Cashiers together earn **$125** per hour." in math notation can be written as:

$$m + 5c = \$125$$

4. The fact "Each Cashier will get a **5**% raise", tells us that the Cashier's new wage per hour in Math notation can be written as:

$$c + 5\% \, c$$

In plain English, this means that the Cashier is currently earning **c** and now is getting plus **5**% of **c**, and the new wage per hour will be:

$$\boldsymbol{c + 5\% \, c}$$

5. The fact "The Manager's raise will be the total of the **5** Cashier raises" tells us that the Manager is very heavily rewarded and his new wage per hour in Math notation can be written as:

$$m + 5 \times 5\% \, c$$

In plain English, this means that the Manager is currently earning **m** and now is getting plus whatever all Cashiers are getting. Since each Cashier is getting **5% c** and there are **5** of them, then the Manager's new wage per hour will be **$m + 5 \times 5\% \, c$**.

6. The fact "Together they will earn **$135** per hour", tells us that the Manager's and **5** Cashier's new wages per hour will total **$135**, and in math notation this can be written as:

$$(m + 5 \times 5\% \, c) + 5 \times (c + 5\% \, c) = \$135$$

Let's fix this last equation by removing the parenthesis and express the percentages as a fraction:

$$m + 5 \times \frac{5}{100}c + 5(c + \frac{5}{100}c) = 135$$

$$m + \frac{25}{100}c + 5c + \frac{25}{100}c = 135$$

Let's multiply every term by **100** so that we can eliminate the denominator:

$$100m + 100 \times \frac{25}{100}c + 500c + 100 \times \frac{25}{100}c = 100 \times 135$$

$$100m + 25c + 500c + 25c = 13{,}500$$

$$100m + 550c = 13{,}500$$

This modified equation represents in fact the new total of manager's and cashiers wages per hour.

Since the above facts are related, then we can use Systems of Equations to find out **"the dollar amount that the Manager and Cashiers are currently being paid per hour"**. So, the following is our System of Equations:

$$\begin{cases} m + 5c = 125 \\ 100m + 550c = 13{,}500 \end{cases}$$

Let's use the **"Substitution Method"** to solve this system of equations:

We need to isolate a variable in one equation, substitute its value in the other equation, and find the numeric value of the variable.

Let's isolate variable **m** in the first equation, which is easy to do:

$$\begin{cases} m = 125 - 5c \\ 100m + 550c = 13{,}500 \end{cases}$$

Let's substitute **m** in the second equation and solve for **c**:

$$100m + 550c = 13{,}500$$

$$100 \times (125 - 5c) + 550c = 13{,}500$$

$$12{,}500 - 500c + 550c = 13{,}500$$

$$-500c + 550c = 13{,}500 - 12{,}500$$

$$50c = 1{,}000$$

$$c = \frac{1{,}000}{50}$$

$$c = 20$$

Now that we found the value of c we can easily find m using the first equation:

$$m = 125 - 5c$$

$$m = 125 - 5 \times 20$$

$$m = 125 - 100$$

$$m = 25$$

So, we found that Cashier is being paid **$20** per hour and the Manager is paid **$25** per hour.

Check your work

Let's use $c = 20$ and $m = 25$ and confirm that our equations are in balance.

Our System of Equations is:

$$\begin{cases} m + 5c = 125 \\ 100m + 550c = 13{,}500 \end{cases}$$

After we plug in the values for x and y:

$$\begin{cases} 25 + 5 \times 20 = 125 \\ 100 \times 25 + 550 \times 20 = 13{,}500 \end{cases}$$

$$\begin{cases} 125 = 125 \\ 13{,}500 = 13{,}500 \end{cases}$$

This proves that our calculations were correct.

If you want to check and confirm their future earnings, then you have to find **5%** increase for one cashier and then multiply it by **5** and add it to the Manager's wage.

For **1** Cashier who gets **5%** raise:

$$5\% \times c = \frac{5}{100} \times 20 = 1$$

So the Cashier's wage increase is **$1**. Since there were **5** Cashiers, and the Manager wage will increase the total of all **5** Cashier raises, then his increase will be **$5**. So the new wages will be **$21** for the Cashier and **$30** for the Manager. Now we can calculate how much they will earn all together:

$$m + 5c = ?$$

$$\$30 + 5 \times \$21 = \$135$$

This confirms our fact 6.

38. Get the Money ready

Let's say that you and your friend went into a Bakery and decided to buy **2** Coffee and **2** Bagels.

As you wait on line to pay, you hear that the girl is charging someone in front of you **$5.55** for **3** coffee and **3** bagels. The next person on line pays **$3.70** for **2** coffee and **4** bagels.

How much you will have to pay?

Solution

In order to find how much you will have to pay, we need to find the prices for Coffee and Bagel. Let's extract the facts that are available to us.

From the wording of our question, it is clear that:

1. There are two types of prices: a price for Coffee and a price for Bagel.
2. " **$5.55** for **3** Coffee and **3** Bagels."
3. " **$5.20** for **2** Coffee and **4** Bagels"

Now, let's "translate" the facts given in sentences into a mathematical notation:

1. Since we need to find out the price for Coffee and the price for Bagel AND they are unknown to us at this moment, we will simply use two math symbols for unknowns such as c and b to mark them. Let's use:
     c - for the Coffee price, and
     b - for the Bagel price.

2. The fact " **$5.55** for **3** Coffee and **3** Bagels." in Math notation can be written as:

$$3c + 3b = \$5.55$$

3. The fact " $\$5.20$ for **2** Coffee and **4** Bagels" in math notation can be written as:

$$2c + 4b = \$5.20$$

Since the above facts are related, then we can use Systems of Equations to find out **"the prices for Coffee and Bagel"**. So, the following is our System of Equations:

$$\begin{cases} 3c + 3b = 5.55 \\ 2c + 4b = 5.20 \end{cases}$$

Let's use the **"Addition Method"** to solve this system of equations:
We need to make the coefficients of one variable in both equations the same but with opposite signs, so that when we add the equations, the variable gets eliminated. Let's try to eliminate variable **c**. One way to achieve this is to multiply the first equation by **2** and the second equation by (-3):

$$\begin{cases} 3c + 3b = 5.55 & /\times 2 \\ 2c + 4b = 5.20 & /\times (-3) \end{cases}$$

$$\begin{cases} 6c + 6b = 11.1 \\ -6c - 12b = -15.6 \end{cases}$$

Let's add the equations side by side

$$(6c + 6b) + (-6c - 12b) = 11.1 + (-15.6)$$

$$6c + 6b - 6c - 12b = 11.1 - 15.6$$

$$-6b = -4.5$$

$$b = \frac{-4.5}{-6}$$

$$b = 0.75$$

Now that we found the value of **b** we can easily find **c** using the first equation:

$$3c + 3b = 5.55$$

$$3c + 3 \times 0.75 = 5.55$$

$$3c + 2.25 = 5.55$$

$$3c = 5.55 - 2.25$$

$$3c = 3.3$$

$$c = \frac{3.3}{3}$$

$$c = 1.1$$

So, we found that the price for Coffee is **$1.1** and the price for Bagel **$0.75**. Since you are getting **2** coffee and **2** bagels, then you will have to pay:

$$2c + 2b = 2 \times 1.1 + 2 \times 0.75 = \$3.70$$

Check your work

Let's use **$c = 1.1$** and **$b = 0.75$** and confirm that our equations are in balance.

Our System of Equations is:

$$\begin{cases} 3c + 3b = \$5.55 \\ 2c + 4b = \$5.20 \end{cases}$$

After we plug in the values for **x** and **y**:

$$\begin{cases} 3 \times \$1.1 + 3 \times \$0.75 = \$5.55 \\ 2 \times \$1.1 + 4 \times \$0.75 = \$5.20 \end{cases}$$

$$\begin{cases} \$5.55 = \$5.55 \\ \$5.20 = \$5.20 \end{cases}$$

This proves that our calculations were correct.

39. Student loans

There are many companies who invest their money by loaning it to students so that they can pay for their education.

Let's say that you got two student loans totaling $\$10,000$, and let's say that throughout the year you don't pay anything and they calculate your interest payment once a year at the end of the year.

At the end of the year they mail you the 1098-E form to inform you that they charged and will collect $\$560$ interest from you and you can deduct this when you do your taxes.

If for your first loan you pay 5% interest, and for the second loan you pay 6.5% interest, then how much is the first loan and how much is the second loan?

Solution

In order to find out how much is your first and second loan, we need to fully understand what is going on here. Let's extract the facts that are available to us.

From the wording of our question, it is clear that:

1. You borrowed money two times and got two separate loans with two different interest rates.
2. "Two student loans totaling $\$10,000$"
3. "For your first loan you pay 5% interest"
4. "For your second loan you pay 6.5% interest"
5. "They charged or they will collect $\$560$ interest"

Now, let's "translate" the facts given in sentences into a mathematical notation:

1. Since we need to find out how much is each of the two student loans AND they are unknown to us at this moment, we will simply use two math symbols for unknowns such as f and s to mark them. Let's use:

f - for the first student loan, and

s - for the second student loan.

2. The fact that the "two student loans total **$10,000**" in Math notation can be written like:

$$f + s = \$10{,}000$$

3. The fact: "for your first loan you pay **5%** interest", in Math notation can be written like:

$$5\%f = \frac{5}{100}f = 0.05\,f$$

4. The fact: "for your second loan you pay **6.5%** interest", in Math notation can be written like:

$$6.5\%s = \frac{6.5}{100}s = 0.065\,s$$

5. The fact: they collected **$560** interest, means that the interest from first loan, plus the interest from the second loan equals **$560**. Here we can use the findings from facts **3** and **4** above and express fact **5** in math notation like:

$$0.05\,f + 0.065\,s = \$560$$

Since the above facts are related, then we can use Systems of Equations to find out **"the amount of the first and second loan you borrowed for your Education"**. So, the following is our System of Equations:

$$\begin{cases} f + s = 10{,}000 \\ 0.05\,f + 0.065\,s = 560 \end{cases}$$

Let's use the **"Addition Method"** to solve this system of equations:

We need to make the coefficients of one variable in both equations the same but with opposite signs, so that when we add the equations, the variable gets eliminated. Let's try to eliminate variable s. One way to achieve this is to multiply the first equation by (-0.05):

$$\begin{cases} f + s = 10{,}000 & /\times(-0.05) \\ 0.05\,f + 0.065\,s = 560 \end{cases}$$

$$\begin{cases} -0.05\,f - 0.05\,s = -500 \\ 0.05\,f + 0.065\,s = 560 \end{cases}$$

Let's add the equations side by side

$$(-0.05\,f - 0.05\,s) + (0.05\,f + 0.065\,s) = (-500) + 560$$
$$-0.05\,f - 0.05\,s + 0.05\,f + 0.065\,s = -500 + 560$$
$$0.015\,s = 60$$
$$s = \frac{60}{0.015}$$
$$s = 4{,}000$$

Now that we found the value of s we can easily find f using the first equation:

$$f + s = 10{,}000$$
$$f + 4{,}000 = 10{,}000$$
$$f = 10{,}000 - 4{,}000$$
$$f = 6{,}000$$

So, we found that the first student loan at **5%** interest is **$6,000** and the second student loan at **6.5%** interest is **$4,000**.

Check your work

Let's use $f = 6,000$ and $s = 4,000$ and confirm that our equations are in balance.

Our System of Equations is:

$$\begin{cases} f + s = 10,000 \\ 0.05\,f + 0.065\,s = 560 \end{cases}$$

After we plug in the values for x and y:

$$\begin{cases} 6,000 + 4,000 = 10,000 \\ 0.05 \times 6,000 + 0.065 \times 4,000 = 560 \end{cases}$$

$$\begin{cases} 10,000 = 10,000 \\ 560 = 560 \end{cases}$$

This proves that our calculations were correct.

40. Study time for SAT

Let's say that you decided to study **250 _min_** every week for SAT. A good part of it you will spend for Math SAT and part for English SAT.

If the ratio of time spent on Math SAT to time spent on English SAT will be **3** to **2**, then how many minutes a week you will be spending on Math SAT and how many minutes on English SAT?

Solution

In order to find out how many minutes a week you will be spending on Math SAT and how many minutes on English SAT, we need to fully understand what is going on here. Let's extract the facts that are available to us.

From the wording of our question, it is clear that:

1. You will study two subjects for SAT: Math and English.
2. "**250 _min_** every week for SAT"
3. "The ratio of time spent on Math SAT to time spent on English SAT will be **3** to **2**"

Now, let's "translate" the facts given in sentences into a mathematical notation:

1. Since we need to find out Math time and English time AND they are unknown to us at this moment, we will simply use two Math symbols for unknowns such as **_m_** and **_e_** to mark them. Let's use:

    _m_ - for Math time in minutes, and

    _e_ - for English time in minutes.
2. The fact "**250 _min_** every week for SAT" means that the time spent for Math and time spent for English for one week will be **250 _min_**, and in Math notation this can be written like:

$$m + e = 250$$

3. The fact: "The ratio of time spent on Math SAT to time spent on English SAT will be **3** to **2**", looks a bit tricky but it is not. Recall that one way to write ratios in Math notation is using Fractions. So, when we say that the ratio of Math time (**m**) to English time (**e**) is **3 to 2** , in Math notation this statement can be written like:

$$\frac{m}{e} = \frac{3}{2}$$

Since the above facts are related, then we can use Systems of Equations to find out **"the minutes you will spend on Math SAT and the minutes you will spend on English SAT"**. So, the following is our System of Equations:

$$\begin{cases} m + e = 250 \\ \dfrac{m}{e} = \dfrac{3}{2} \end{cases}$$

Let's use the **"Substitution Method"** to solve this system of equations:

We need to isolate one variable in one equation and substitute it in the other equation. Let's isolate **m** in the second equation:

$$\begin{cases} m + e = 250 \\ m = \dfrac{3}{2}e \end{cases}$$

Let's substitute the value of **m** of the second equation in the first equation. So, we have:

$$m + e = 250$$

$$\frac{3}{2}e + e = 250$$

Let's multiply every term by **2**, so that we don't have to work with fractions:

$$2 \times \frac{3}{2}e + 2 \times e = 2 \times 250$$

$$3e + 2e = 500$$

$$5e = 500$$

$$e = \frac{500}{5}$$

$$e = 100$$

Now that we found the value of **e** we can easily find **m** using the first equation:

$$m + e = 250$$

$$m + 100 = 250$$

$$m = 250 - 100$$

$$m = 150$$

So, we found that you will study **150 min** a week Math SAT, and **100 min** English SAT.

Check your work

Let's use **m = 150** and **e = 100** and confirm that our equations are in balance.

Our System of Equations is:

$$\begin{cases} m + e = 250 \\ \dfrac{m}{e} = \dfrac{3}{2} \end{cases}$$

After we plug in the values for **x** and **y**:

$$\begin{cases} 150 + 100 = 250 \\ \dfrac{150}{100} = \dfrac{3}{2} \end{cases}$$

$$\begin{cases} 250 = 250 \\ \dfrac{3 \times 50}{2 \times 50} = \dfrac{3}{2} \end{cases}$$

$$\begin{cases} 250 = 250 \\ \dfrac{3}{2} = \dfrac{3}{2} \end{cases}$$

This proves that our calculations were correct.

41. Basketball shots

Let's say that during a basketball game you scored **27** points.

You threw **12** shots and no free throws.

Find out how many two point-shots and how many three-point shots you scored?

Solution

Let's extract the facts that are available to us.

From the question, it is clear that:

1. There are two types of basketball shots: two-pointers and three-pointers
2. "you scored **27** points"
3. "You threw **12** shots and no free throws."

Now, let's "translate" the facts given in sentences into a mathematical notation:

1. Since we need to find out two-pointers and three-pointers AND they are unknown to us at this moment, we will simply use two Math symbols for unknowns such as x and y to mark them. Let's use:

    x - For three-pointers, and

    y - For two-pointers.

2. The fact "you scored **27** points" means that when we add the two-pointers and three-pointers the result is **27**, and in math notation this can be written like:

$$3x + 2y = 27$$

3. The fact: "You threw **12** shots and no free throws."", means that since there were no free throws (one-pointers) then all shots were either two or three pointers, and in math notation this statement about the number of shots can be written like:

$$x + y = 12$$

Since the above facts are related, then we can use Systems of Equations to find out **"how many two-point shots and three-point shots you scored"**. So, the following is our System of Equations:

$$\begin{cases} 3x + 2y = 27 \\ x + y = 12 \end{cases}$$

Let's use the **"Substitution Method"** to solve this system of equations:

We need to isolate one variable in one equation and substitute it in the other equation. Let's isolate x in the second equation:

$$\begin{cases} 3x + 2y = 27 \\ x = 12 - y \end{cases}$$

Let's substitute the value of **x** of the second equation in the first equation. So, we have:

$$3x + 2y = 27$$
$$3(12 - y) + 2y = 27$$
$$36 - 3y + 2y = 27$$
$$36 - y = 27$$
$$-y = 27 - 36$$
$$-y = -9$$
$$y = 9$$

Now that we found the value of y we can easily find x using the second equation:

$$x + y = 12$$
$$x + 9 = 12$$
$$x = 12 - 9$$
$$x = 3$$

So, we found that you scored **9** two-point shots and **3** three-point shots.

Check your work

Let's use $x = 3$ and $y = 9$ and confirm that our equations are in balance.

Our System of Equations is:

$$\begin{cases} 3x + 2y = 27 \\ x + y = 12 \end{cases}$$

After we plug in the values for x and y:

$$\begin{cases} 3 \times 3 + 2 \times 9 = 27 \\ 3 + 9 = 12 \end{cases}$$

$$\begin{cases} 27 = 27 \\ 12 = 12 \end{cases}$$

This proves that our calculations were correct.

42. Basketball shooting competition

Let's say that you and a friend of yours are having a basketball three-point shooting competition.

The rules are as follows: if you miss a shot, you lose **5** points, otherwise you earn **3** points.

It happened that your **40**th shot was a miss and you had no points left.

Find out how many shots you missed and how many you made?

Solution

Let's extract the facts that are available to us.

From the question, it is clear that:

1. There are two types of basketball shots: the made ones and missed ones.
2. "Miss a shot, you lose **5** points"
3. "Make a shot, you earn **3** points"
4. "You threw **40** shots"
5. "You have no points left"

Now, let's "translate" the facts given in sentences into a mathematical notation:

1. Since we need to find out made-shots and missed-shots AND they are unknown to us at this moment, we will simply use two math symbols for unknowns such as x and y to mark them. Let's use:

    x - For made-shots, and

    y - For missed-shots.

2. The fact "miss a shot, you lose **5** points" means that for each missed shot y, we have to multiply it by **5**, and because we lose points, we have to use a negative $(-)$ sign. So, in math notation this fact can be written like:

$$-5y$$

3. The fact "make a shot, you earn **3** points" means that for each made shot x, we have to multiply it by **3**, and in math notation this fact can be written like:

$$3x$$

4. The fact "You threw **40** shots" means that the combination of the shots made and missed is **40**, and in math notation this fact can be written like:

$$x + y = 40$$

5. The fact "You have no points left" means that the combination of the points earned and lost after x and y shots is **0**. We already discussed in point **2** and **3** above how to write the points, and those findings we can express in Math notation like:

$$3x - 5y = 0$$

Since the above facts are related, then we can use Systems of Equations to find out **"how many shots you missed and how many you made"**. So, the following is our System of Equations:

$$\begin{cases} x + y = 40 \\ 3x - 5y = 0 \end{cases}$$

Let's use the **"Substitution Method"** to solve this system of equations:

We need to isolate one variable in one equation and substitute it in the other equation. Let's isolate x in the first equation:

$$\begin{cases} x = 40 - y \\ 3x - 5y = 0 \end{cases}$$

Let's substitute the value of x of the first equation in the second equation. So, we have:

$$3(40 - y) - 5y = 0$$
$$120 - 3y - 5y = 0$$
$$-8y = -120$$

$$y = \frac{-120}{-8}$$

$$y = 15$$

Now that we found the value of **y** we can easily find **x** using the first equation:

$$x + y = 40$$

$$x + 15 = 40$$

$$x = 40 - 15$$

$$x = 25$$

So, we found that you made **25** shots and you missed **15** shots.

Check your work

Let's use $x = 2$ and $y = 15$ and confirm that our equations are in balance. Our System of Equations is:

$$\begin{cases} x + y = 40 \\ 3x - 5y = 0 \end{cases}$$

After we plug in the values for **x** and **y**:

$$\begin{cases} 25 + 15 = 40 \\ 3 \times 25 - 5 \times 15 = 0 \end{cases}$$

$$\begin{cases} 40 = 40 \\ 0 = 0 \end{cases}$$

This proves that our calculations were correct.

43. Basketball free-throws

Let's say that you scored **68** points on the last basketball game you played for your high school. In order to achieve this result you made **40** shots, including free-throws, two-pointers and three-pointers.

You made two times as many two-point shots as three-point shots.

Find out how many free-throws did you make?

Solution

It should be clear that even though we are asked to find only the number of free-throws, we will need to find out the number of other types of shots as well.

Let's extract the facts that are available to us from the question:

1. There are three types of basketball shots: free-throws, two-point and three-point shots.
2. "You scored **68** points"
3. "You made **40** shots"
4. "You made two times as many two-point shots as three-point shots."

Now, let's "translate" the facts given in sentences into a mathematical notation:

1. Since we need to find out free-throws, two-point shots and three-point shots AND they are unknown to us at this moment, we will simply use three math symbols for unknowns such as x, y and z to mark them. Let's use:

    x - For free-throws,

    y - For two-point shots and

    z - For three-point shots.

2. The fact "you scored **68** points" means that the number of free-throws times **1** point + the number of two-point shots times **2** points + the number of

three-point shots times **3** points equals **68**. So, in Math notation this fact can be written like:

$$x + 2y + 3z = 68$$

3. The fact "You made **40** shots" means that the number of free-throws + two-point shots + three-point shots equals **40**, and in Math notation this fact can be written like:

$$x + y + z = 40$$

4. The fact "You made two times as many two-point shots as three-point shots" means that the numbers of the two-point shots is **2** times bigger than the number of three-point shots, and in Math notation this fact can be written like:

$$y = 2z$$

Since the above facts are related, then we can use Systems of Equations to find out **"how many free-throws did you make"**. So, the following is our System of Equations:

$$\begin{cases} x + 2y + 3z = 68 \\ x + y + z = 40 \\ y = 2z \end{cases}$$

Let's use the **"Substitution Method"** to solve this system of equations:

We need to isolate one variable in one equation and substitute it in the other equations. Variable **y** is already isolated, and let's substitute it in the other two equations:

$$\begin{cases} x + 2 \times 2z + 3z = 68 \\ x + 2z + z = 40 \end{cases}$$

$$\begin{cases} x + 4z + 3z = 68 \\ x + 3z = 40 \end{cases}$$

$$\begin{cases} x + 7z = 68 \\ x + 3z = 40 \end{cases}$$

Let's use again the **"Substitution Method"** to solve this system of equations. Let's isolate x in the second equation:

$$\begin{cases} x + 7z = 68 \\ x = 40 - 3z \end{cases}$$

Let's substitute its value in the first equation. So, we have:

$$x + 7z = 68$$
$$(40 - 3z) + 7z = 68$$
$$-3z + 7z = 68 - 40$$
$$4z = 28$$
$$z = \frac{28}{4}$$
$$z = 7$$

Now that we found the value of z we can easily find y using the third equation:

$$y = 2z$$
$$y = 2 \times 7$$
$$y = 14$$

Now that we have the values of y and z we can use the second equation to find x:

$$x + y + z = 40$$
$$x + 14 + 7 = 40$$
$$x = 40 - 14 - 7$$
$$x = 19$$

So, we found that you made **19** free-throws, **14** two-point shots and **7** three-point shots.

Check your work

Let's use $x = 19$, $y = 14$ and $z = 7$, and confirm that our equations are in balance.

Our System of Equations is:

$$\begin{cases} x + 2y + 3z = 68 \\ x + y + z = 40 \\ y = 2z \end{cases}$$

After we plug in the values for x and y:

$$\begin{cases} 19 + 2 \times 14 + 3 \times 7 = 68 \\ 19 + 14 + 7 = 40 \\ 14 = 2 \times 7 \end{cases}$$

$$\begin{cases} 68 = 68 \\ 40 = 40 \\ 14 = 14 \end{cases}$$

This proves that our calculations were correct.

44. Number of Coins in pocket

Let's say that Artie has **25** coins in his pocket, worth **$4.15**, all in quarters and dimes.

Calculate how many quarters and dimes Artie has in his pocket?

Solution

Let us write down the facts that are available to us from the question:

1. Artie has two types of coins: quarters and dimes.
2. "Artie has **25** coins"
3. "**25** coins are worth **$4.15**"

Now, let's "translate" the facts given in sentences into a mathematical notation:

1. Since we need to find out the number of quarters and the number of dimes AND they are unknown to us at this moment, we will simply use two math symbols for unknowns such as *q* and *d* to mark them. Let's use:
    q - For quarters, and
    d - For dimes.
2. The fact "Artie has **25** coins" means that the number of quarters + the number of dimes equals **25**. So, in Math notation this fact can be written like:

$$q + d = 25$$

3. The fact "coins are worth **$4.15**" means that the value of quarters + the value of dimes equals **$4.15**. The value of a quarter, as you know, is **$0.25** and the value of a dime is **$0.10**. When these values are multiplied by the number of coins such as *q* and *d* then we will get the total dollar value of all coins. Hence this fact in math notation can be written like:

$$0.25q + 0.10d = 4.15$$

It is obvious that using the Systems of Equations we will be able to find out q and d in the above equations. So, the following is our System of Equations:

$$\begin{cases} q + d = 25 \\ 0.25q + 0.10d = 4.15 \end{cases}$$

Let's use the **"Substitution Method"** to solve this system of equations:

We need to isolate one variable in one equation and substitute it in the other equations. Let's isolate q:

$$\begin{cases} q = 25 - d \\ 0.25q + 0.10d = 4.15 \end{cases}$$

Let's substitute q in the second equation and solve for d:

$$0.25q + 0.10d = 4.15$$
$$0.25 \times (25 - d) + 0.10d = 4.15$$
$$6.25 - 0.25d + 0.10d = 4.15$$
$$-0.15d = 4.15 - 6.25$$
$$-0.15d = -2.10$$
$$d = \frac{-2.10}{-0.15}$$
$$d = 14$$

Now that we found the number of dimes d we can easily find q using the first equation:

$$q = 25 - d$$
$$q = 25 - 14$$
$$q = 11$$

So, we found that Artie has **11** quarters and **14** dimes in his pocket.

Check your work

Let's use $q = 11$ and $d = 14$ and confirm that our equations are in balance.

Our System of Equations is:

$$\begin{cases} q + d = 25 \\ 0.25q + 0.10d = 4.15 \end{cases}$$

After we plug in the values for x and y:

$$\begin{cases} 11 + 14 = 25 \\ 0.25 \times 11 + 0.10 \times 14 = 4.15 \end{cases}$$

$$\begin{cases} 25 = 25 \\ 4.15 = 4.15 \end{cases}$$

This proves that our calculations were correct.

45. Number of Coins in wallet

Let's say that Nina has **21** coins in her wallet, worth **$3.25**, all in quarters, dimes and nickels.

If the value of dimes is **4** times as much as that of nickels, then how many quarters, dimes and nickels Nina has in her wallet?

Solution

Let us write down the facts that are available to us from the question:

1. Nina has three types of coins: quarters, dimes and nickels.
2. "Nina has **21** coins"
3. " **21** Coins are worth **$3.25**"
4. "The value of dimes is **4** times that of nickels"

Now, let's "translate" the facts given in sentences into a mathematical notation:

1. Since we need to find out the number of quarters, the number of dimes and the number of nickels AND they are unknown to us at this moment, we will simply use three math symbols for unknowns such as q, d and n to mark them. Let's use:
   q - For quarters,
   d - For dimes and
   n - For nickels.
2. The fact "Nina has **21** coins" means that the number of quarters + the number of dimes + the number of nickels equals **21**. So, in Math notation this fact can be written like:

$$q + d + n = 21$$

3. The fact " **21** coins are worth **$3.25**" means that the value of quarters + the value of dimes + the value of nickels equals **$3.25**. The value of a quarter, as you know,

is $0.25 and the value of a dime is $0.10 and that of a nickel is $0.05. When these values are multiplied by the number of coins such as q, d and n then we will get the total dollar value of all coins. Hence this fact in Math notation can be written like:

$$0.25\,q + 0.10\,d + 0.05\,n = 3.25$$

4. The fact "The value of dimes is **4** times that of nickels" in math notation can be written like:

$$0.10\,d = 4 \times 0.05\,n$$

Using the Systems of Equations we will be able to find out q, d and n in the above equations. So, the following is our System of Equations:

$$\begin{cases} q + d + n = 21 \\ 0.25\,q + 0.10\,d + 0.05\,n = 3.25 \\ 0.10\,d = 4 \times 0.05\,n \end{cases}$$

Since the third equation has only two variables, let's isolate d and then substitute its value in the other equations:

$$0.10\,d = 4 \times 0.05\,n$$
$$0.10\,d = 0.2\,n$$
$$d = \frac{0.2}{0.10}n$$
$$d = 2n$$

Now the system of equations looks simpler:

$$\begin{cases} q + d + n = 21 \\ 0.25q + 0.10d + 0.05n = 3.25 \\ d = 2n \end{cases}$$

Let's substitute the value of d of the third equation in the other two:

$$\begin{cases} q + 2n + n = 21 \\ 0.25q + 0.10 \times 2n + 0.05n = 3.25 \end{cases}$$

$$\begin{cases} q + 3n = 21 \\ 0.25q + 0.2n + 0.05n = 3.25 \end{cases}$$

$$\begin{cases} q + 3n = 21 \\ 0.25q + 0.25n = 3.25 \end{cases}$$

Let's use the **"Substitution Method"** to solve this system of equations:

We need to isolate one variable in one equation and substitute it in the other equations. Let's isolate q:

$$\begin{cases} q = 21 - 3n \\ 0.25q + 0.25n = 3.25 \end{cases}$$

Let's substitute q in the second equation and solve for n:

$$0.25q + 0.25n = 3.25$$
$$0.25 \times (21 - 3n) + 0.25n = 3.25$$
$$5.25 - 0.75n + 0.25n = 3.25$$
$$-0.75n + 0.25n = 3.25 - 5.25$$
$$-0.5n = -2$$
$$n = \frac{-2}{-0.5}$$
$$n = 4$$

Now that we found the number of nickels n we can easily find d using the third equation:

$$d = 2n$$
$$d = 2 \times 4$$
$$d = 8$$

Using the first equation we can find the value of q:

$$q + d + n = 21$$
$$q + 8 + 4 = 21$$
$$q = 21 - 12$$
$$q = 9$$

So, we found that Nina has **9** quarters, **8** dimes and **4** nickels in her wallet.

Check your work

Let's use $q = 9$, $d = 8$ and $n = 4$, and confirm that our equations are in balance. Our System of Equations is:

$$\begin{cases} q + d + n = 21 \\ 0.25q + 0.10d + 0.05n = 3.25 \\ 0.10d = 4 \times 0.05n \end{cases}$$

After we plug in the values for x and y:

$$\begin{cases} 9 + 8 + 4 = 21 \\ 0.25 \times 9 + 0.10 \times 8 + 0.05 \times 4 = 3.25 \\ 0.10 \times 8 = 4 \times 0.05 \times 4 \end{cases}$$

$$\begin{cases} 21 = 21 \\ 3.25 = 3.25 \\ 0.8 = 0.8 \end{cases}$$

This proves that our calculations were correct.

46. Legs of a Triangle

Let's say that the area of a right angled triangle is **24 *sq inches***.

If the ratio of the lengths of its legs is **1 : 3 (1 to 3)**, then what are the lengths of its legs?

Solution

Let us write down the two facts that are available to us from the question:

1. Area of a right angled triangle is **24 *sq inches***.
2. "The ratio of its legs is **1 : 3**"

It is very important, before we do anything else, to make sure that we understand the problem. In a few words, we have the Area right angled triangle and the ratio of the legs **a** and **b**, and we need to find their lengths. The first thing you should recall is the formula for calculating the Area of a right angled triangle, which is:

$$A = \frac{a \times b}{2}$$

where **a** and **b** are the legs, and **A** the area.

Since in our case **A = 24**, then we can say:

$$\frac{a \times b}{2} = 24$$

When it is said: the ratio of its legs is **1 : 3**, in Math notation this can be written like:

$$\frac{a}{b} = \frac{1}{3}$$

In order to find the lengths of the legs, we can use the above two facts and construct a System of Equations and find out the length of the legs of the right angled triangle. So, the following is our System of Equations:

$$\begin{cases} \dfrac{a \times b}{2} = 24 \\ \dfrac{a}{b} = \dfrac{1}{3} \end{cases}$$

Let us fix and make the Equations simpler:

$$\begin{cases} a \times b = 24 \times 2 \\ 3\dfrac{a}{b} = 1 \end{cases}$$

$$\begin{cases} ab = 48 \\ 3a = b \end{cases}$$

If we were to use the **Substitution Method** and substitute the value of b from the second equation in the first equation, then we would be able to calculate a. So we have:

$$ab = 48$$

$$a \times 3a = 48$$

$$3a^2 = 48$$

$$a^2 = \frac{48}{3}$$

$$a^2 = 16$$

$$a = \pm\sqrt{16}$$

$$a = \pm 4$$

You understand that $\pm$ is there because Quadratic Equations have two solutions: a positive and a negative solution. However, not always both solutions make sense in the Real world.

In our case, to say the length of one of the legs of a right angled triangle is -4 does not make sense. As you know, the length cannot be negative, hence we will ignore this value and stick to $a = 4$ only.

Now that we found a we can easily find b using the second equation:

$$3a = b$$

$$3 \times 4 = b$$

$$b = 12$$

So, we found that the legs (sides) of the right angled triangle are: $a = 4$ and $b = 12$.

Check your work

Let's use $a = 4$ and $b = 12$ and confirm that our equations are in balance.

Our System of Equations is:

$$\begin{cases} \dfrac{a \times b}{2} = 24 \\ \dfrac{a}{b} = \dfrac{1}{3} \end{cases}$$

After we plug in the values for x and y:

$$\begin{cases} \dfrac{4 \times 12}{2} = 24 \\ \dfrac{4}{12} = \dfrac{1}{3} \end{cases}$$

$$\begin{cases} \dfrac{48}{2} = 24 \\[2ex] \dfrac{1 \times 4}{3 \times 4} = \dfrac{1}{3} \end{cases}$$

$$\begin{cases} 24 = 24 \\[2ex] \dfrac{1}{3} = \dfrac{1}{3} \end{cases}$$

This proves that our calculations were correct.

47. Shipment Quantity

Imagine you own a business where you sell sport items. Let's say that you receive a shipment of Nike and Adidas sneakers.

By the end of the week you calculate that you sold Nike and Adidas sneakers in the amount of $\$\,6,450$. You count the inventory and you find that you still did not sell 25% of Nike sneakers and 75% Adidas sneakers that you received last week. Once you sell these too, then you will generate additional $\$\,5,350$.

If Nike sneakers sell for $\$\,70$ and Adidas sneakers sell for $\$\,60$, then how many Nike sneakers and how many Adidas sneakers did you receive in the last week's shipment?

Solution

As we explained in the previous examples, understanding the facts and details given in the question is of utmost importance. This exercise is fairly difficult one and contains many hidden details, so pay attention as we explain them.

From the wording of our question, we understand that there are three facts given to us:

1. "Nike and Adidas sneakers shipped" is one fact. We don't know how many were shipped, and that is what we need to find out.
2. "Nike and Adidas sneakers sold in the amount of $\$\,6,450$". This is another fact which is given to us. Question: can we tell how many Nike and how many Adidas sneakers were sold to generate $\$\,6,450$ in sales? Answer: Yes we can. In fact, we cannot tell in exact absolute numbers until we solve the problem, but from the fact **3** below we can easily figure out the portions in relative numbers (percentages) that generated the sales, as you will see below.
3. "**25** % of Nike sneakers and **75**% Adidas sneakers will generate $\$5,350$". This is the third fact given to us explicitly and it needs a little bit of explanation. These are the percentages of sneakers which **were not** sold and are still in the store. From this explicit fact we can derive another fact, an implicit fact, which is fact number **4**:

4. " **75** % of the Nike and **25%** of Adidas sneakers" were already sold. This is the fourth fact which is given to us implicitly, which means that it is not given to us in the wording of the question directly but we can understand indirectly or derive it from the fact **3** above.

5. "Nike sneakers sell for **$70** and Adidas sneakers sell for **$60**". This is the fifth fact which we will use to solve the problem at the end.

Wow, these were a lot of details to digest. Take a deep breath and continue reading.

Now, let's "translate" the facts given in sentences into a mathematical notation:

1. "Nike and Adidas sneakers shipped". We don't know how many sneakers but we know the total dollar amount, and we need to distinguish or separate the dollar amount for Nike and the dollar amount for Adidas sneakers. Since we need first to find out how much (in dollar terms) of these items were shipped to us AND they are unknown to us at this moment, we will simply use two math symbols for unknowns such as **x** and **y**. Let's use:

   - **x** for the dollar amount of Nike sneakers shipped, and
   - **y** for the dollar amount of Adidas sneakers shipped.

Remember that **x** and **y** do not represent the number of Nike and Adidas sneakers, but **x** and **y** represent the DOLLAR value of the shipment. This is so because the other facts show in dollars what is sold and what is still to be sold.

2. "Nike and Adidas sneakers sold in the amount of **$6,450**". In order to write this fact in mathematical notation, we need to know how much from the shipment we sold. Note that it would be incorrect to write:

$$\mathbf{x} + \mathbf{y} = \$6,450 \quad \text{-- wrong!}$$

This is because we said that **x** and **y** represent the dollar amount what was **shipped** and not what was sold. So, we need to combine this fact with the implicit fact number **4** above: " **75%** of Nike sneakers and **25%** of Adidas were sold". If we use **x** and **y**, then the facts number **2** and **4** combined, in math notation can be written like:

$$75\% \, x + 25\% \, y = \$6{,}450$$

Let's express percentages as fractions and fix (simplify) the equation:

$$\frac{75}{100}x + \frac{25}{100}y = \$6{,}450$$

$$\frac{3}{4}x + \frac{1}{4}y = \$6{,}450$$

3. **25%** of Nike sneakers and **75%** Adidas sneakers will generate **$5, 350**". This fact tells us the value of unsold Nike and Adidas sneakers. In math notation, this fact can be written like:

$$25\% \, x + 75\% \, y = \$5{,}350$$

Let's express percentages as fractions and fix (simplify) the equation:

$$\frac{25}{100}x + \frac{75}{100}y = \$5{,}350$$

$$\frac{1}{4}x + \frac{3}{4}y = \$5{,}350$$

Since the above facts are related, then we can use the Systems of Equations to find out the dollar amount of Nike and the dollar amount Adidas sneakers shipped. So, the following is our system of equations:

$$\begin{cases} \dfrac{3}{4}x + \dfrac{1}{4}y = \$6{,}450 \\ \dfrac{1}{4}x + \dfrac{3}{4}y = \$5{,}350 \end{cases}$$

It is preferable to eliminate fractions if we can, because it is simpler and easier to calculate. It is obvious that if we multiply all the terms in both equations by **4**, then our system of equations will be simpler:

$$\begin{cases} \dfrac{3}{4}x + \dfrac{1}{4}y = \$6{,}450 & / \times 4 \\[2mm] \dfrac{1}{4}x + \dfrac{3}{4}y = \$5{,}350 & / \times 4 \end{cases}$$

$$\begin{cases} 3x + y = \$25{,}800 \\ x + 3y = \$21{,}400 \end{cases}$$

Let's solve our System of Equations.

We will use the **"Addition Method"** to solve our System of Equations. Recall from previous examples that the purpose of the Addition method is to eliminate one of the variables when we ADD the sides of the two equations. Before we do the addition, we need to have the coefficients of one of the variables to have the same value but opposite signs. Let's decide to eliminate first the **x** variable. In order to do that we can multiply the second equation by (-3):

$$\begin{cases} 3x + y = \$25{,}800 \\ x + 3y = \$21{,}400 & / \times (-3) \end{cases}$$

$$\begin{cases} 3x + y = 25{,}800 \\ -3x - 9y = -64{,}200 \end{cases}$$

Now we add the equations (side by side):
$$(3x + y) + (-3x - 9y) = 25800 + (-64200)$$
$$3x + y - 3x - 9y = 25800 - 64200$$

The variable **x** gets eliminated when we add the like terms:
$$y - 9y = 25800 - 64200$$
$$-8y = -38400$$
$$y = \frac{-38400}{-8}$$
$$y = 4800$$

We will now use the value of **y** to find **x**. Let's use the second equation to find the value of **x**:

$$x + 3y = 21400$$
$$x = 21400 - 3y$$

Let's substitute **y** with its value:

$$x = 21400 - 3 \times 4800$$

$$x = 21400 - 14400$$

$$x = 7000$$

Since we used **x** for the dollar amount of Nike sneakers shipped and **y** for the dollar amount of Adidas sneakers shipped, then **x = 7000** means that the value of Nike sneakers shipped is **$7,000** and **y = 4800** means that the value of Adidas sneakers shipped is **$4,800**.

We found the dollar amount of shipments. However, the question was to find **how many** Nike and **how many** Adidas sneakers were shipped to us, and not the dollar amount? Well, this will be easy because now we have the total dollar value of each type of sneakers, and the price of each pair was already given to us: **$70** for Nike pair, and **$60** for Adidas pair. All we need to do now is to divide:

$$\text{Number\_of\_Sneakers} = \frac{\text{Total\_Amount}}{\text{Price\_Of\_Pair}}$$

Let's find out:

$$\text{Nike} = \frac{7,000}{70} = 100$$

$$\text{Adidas} = \frac{4,800}{60} = 80$$

So, in the last week's shipment there were:

- **100** pairs of Nike sneakers, and
- **80** pairs of Adidas sneakers.

Check your work

Let's use **x = 7000** and **y = 4800** and confirm that our equations are in balance. Our simplified System of Equations was:

$$\begin{cases} \dfrac{3}{4}x + \dfrac{1}{4}y = \$6{,}450 \\ \dfrac{1}{4}x + \dfrac{3}{4}y = \$5{,}350 \end{cases}$$

After we plug in the values for **x** and **y**:

$$\begin{cases} \dfrac{3}{4} \times \$7{,}000 + \dfrac{1}{4} \times \$4{,}800 = \$6{,}450 \\ \dfrac{1}{4} \times \$7{,}000 + \dfrac{3}{4} \times \$4{,}800 = \$5{,}350 \end{cases}$$

$$\begin{cases} \$5{,}250 + \$1{,}200 = \$6{,}450 \\ \$1{,}750 + \$3{,}600 = \$5{,}350 \end{cases}$$

$$\begin{cases} \$6{,}450 = \$6{,}450 \\ \$5{,}350 = \$5{,}350 \end{cases}$$

This proves that our calculations were correct.

48. Business Break Even Point

Let's say that you decided to open up a home business and sell T-Shirts online.

You buy a bunch of **100** T-Shirts at a wholesale price of **$400**. The equipment for printing pictures on a T-Shirt costs **$960**. Each picture you will print on T-Shirt will cost **$2**. The selling price of a T-Shirt will be **$18**.

How many T-Shirts you will need to sell to break even?

Solution

Let's write down the facts that are available to us from the question:

1. You buy **100** T-Shirts for **$400**.
2. Equipment costs **$960**.
3. Each picture costs **$2**.
4. You will sell each T-Shirt for **$18**.

It is very important, not to rush and try to solve the problem. But before you do anything else, you must make sure that you understand the problem in depth. In a few words, every time we are involved in a business there is a Cost associated with it called a Business Expense, and this is the amount of money that you will spend to start and stay in business. The reason why you are in business is because you want to make some extra money after covering the expenses. In your business, you will make money only if you sell a T-Shirt more than it costs to make it, and this is called Profit. At the beginning the costs are high because of Equipment purchase etc. But there is a point where after selling enough T-Shirts, the profit will be equal to the business expense, and this is called "Break Even Point (BEP)". In our case we need to find out how many T-Shirts we will need to sell so that our profit will be equal to **$960**.

The facts that we enumerated above, do not necessarily translate to an equation that we can use to answer our question. We need to look at them and use them to support our "hidden" facts or equations. What do we mean by hidden? Well, we mean facts that are not mentioned in the question. Let's see:

Since the question is "How many T-Shirts to break even?", then we need to find out the profit from selling a single T-Shirt. In order to do this, we need to find out how much does a single T-Shirt will cost.

Since we buy **100** T-Shirts for **$400**, then **1** T-Shirt will cost:

$$1 \ T\_Shirt = \frac{\$400}{100} = \$4$$

Since we need to add a picture to it, which costs **$2**, then the cost of **1** T-Shirt will increase to:

$$1 \ T\_Shirt = \$4 + \$2 = \$6$$

At this moment we are not considering or adding anything else to the cost of T-Shirt such as the Cost of Equipment etc.

Now, the profit **p** from a single T-Shirt will be:

$$p = Selling \ Price - Cost$$

Since our selling price is **$18** and the cost is **$6**, then the profit from a single T-Shirt will be:

$$p = \$18 - \$6$$
$$p = \$12$$

Our profit of **$12** from a single T-Shirt is fixed, and in order to "break even" or to make the money that we invested in equipment we need to sell more T-Shirts so that our total profit be equal to **$960**.

If we use **x** to mark the number of T-Shirts we need to sell, then the following equation shows the braking even model for our T-Shirt business:

$$xp = \$960$$

Since our profit from a single T-Shirt is related to the overall profitability of our business, then we can use the above equations to make a System of equations:

$$\begin{cases} p = 12 \\ xp = 960 \end{cases}$$

If we substitute the value of p of the first equation in the second equation, then we would be able to calculate x. So we have:

$$\begin{array}{cc} xp & 960 \end{array}$$

$$x \times 12 = 960$$

$$x = \frac{960}{12}$$

$$x = 80$$

So, we found we need to sell **80** T-Shirts to break even. Whatever you sell after this, as long as the equipment works, there will be a pure profit for you.

Check your work

Let's use $x = 80$ and $p = 12$ and confirm that our equations are in balance.

Our System of Equations is:

$$\begin{cases} p = 12 \\ xp = 960 \end{cases}$$

After we plug in the values for x and y:

$$\begin{cases} 12 = 12 \\ 80 \times 12 = 960 \end{cases}$$

$$\begin{cases} 12 = 12 \\ 960 = 960 \end{cases}$$

This proves that our calculations were correct.

49. How to choose a phone carrier

Let's say that you decided to get a new smart phone and are considering switching the carrier. You researched the market and are debating whether Company A or Company B is a better deal for you.

Lett's say that Company A is charging **$40** a month for one line with unlimited talk, text and data, but you will have to pay **$750** for your favorite smart phone. Let's say that Company B is charging **$65** a month for one line with unlimited talk, text and data, but it gives you a discount on the phone and you will pay only **$350** for the same favorite smart phone.

After how many months the total cost with either Company A or Company B will be the same? How much that would be?

Solution

Let us write down the facts that are available to us from the question:

1. Company A is charging **$40** a month.
2. Company A is charging **$750** for the phone.
3. Company B is charging **$65** a month.
4. Company B is charging **$350** for the phone.

Before we do anything else, let us digest the facts and understand the problem we are trying to solve. It is clear that Company A is charging less per month than Company B, but it does not give discount for the phone. It is not hard to imagine that over time Company A will turn out to be a better deal, because you are paying less per month. But how long it is going to take? This is exactly what we are trying to find out. The deal with Company A is more expensive at the beginning, and at one point it will be the same as the deal ith Company B and then it will turn to a better deal.

There are two things to observe here:

    a) The number of months you are going to use the phone, and
    b) The total cost up to a particular month.

Let's use:

- x - for the number of months, and
- y - for the total cost.

Let's "translate" the facts that we enumerated above into a mathematical notation:

1. If Company A is charging **$40** a month, and if we want to know how much we paid in total for x number of months, then we would use the following expression to calculate that:

$$40x$$

2. Since Company A is charging **$750** for the phone only one time when you sign up for their service, then this will be added to the total cost y, which will be:

$$y = 40x + 750$$

This model or equation represents the case with Company A if we need to find the total cost y after x number of months.

Similarly for Company B:

3. If Company B is charging **$65** a month, and if we want to know how much we paid in total for x number of months, then we would use the following expression to calculate that:
$$65x$$

4. Since Company B is charging **$350** for the phone only one time when you sign up for their service, then this will be added to the total cost y, which will be:

$$y = 65x + 350$$

Since at some point in time the cost with either carrier will be the same, which means that the x and y in both equations will have the same value, then we can use the Systems of Equations to find those values. Here is our System of Equations:

$$\begin{cases} y = 40x + 750 \\ y = 65x + 350 \end{cases}$$

This will be easy to solve. We can use the value of y of the first equation and substitute it in the second equation. So the second equation is:

$$y = 65x + 350$$

After substitution:

$$40x + 750 = 65x + 350$$

Let's solve for x:

$$40x - 65x = 350 - 750$$
$$-25x = -400$$
$$x = \frac{-400}{-25}$$
$$x = 16$$

Let's substitute the value of x in the second equation to find the value of y:

$$y = 65x + 350$$
$$y = 65 \times 16 + 350$$
$$y = 1390$$

So, we found that it is going to take **16** months for the total cost to be the same whether you sign up with Company A or Company B, and it will be **$1390**.

After **16** months Company A will turn to be cheaper and better choice.

Check your work

Let's use $x = 16$ and $y = 1390$ and confirm that our equations are in balance.

Our System of Equations is:

$$\begin{cases} y = 40x + 750 \\ y = 65x + 350 \end{cases}$$

After we plug in the values for x and y:

$$\begin{cases} 1390 = 40 \times 16 + 750 \\ 1390 = 65 \times 16 + 350 \end{cases}$$

$$\begin{cases} 1390 = 1390 \\ 1390 = 1390 \end{cases}$$

This proves that our calculations were correct.

50. Pizza with toppings

Let's say that Nina and Artie decided to get two large Pizzas for a family dinner. The problem is that Nina and Artie don't have the same taste and always argue which pizza to buy. Nina went to get her favorite Cheese Crust Pizza from Pizzeria Roma, and Artie went to get his Pizza from Pizzeria Tetova.

Pizzeria Roma charges **$14** for a large plain Pizza, and **$1** for each topping.

Pizzeria Tetova charges **$12** for a large plain Pizza, and **$1.50** for each topping.

When Nina and Artie returned home, they both had spent the same amount of money for their Pizza with toppings, and they both got the same number of toppings on their Pizza.

How much money did each Pizza cost, and how many toppings there were on a Pizza?

Solution

As always, first we write down the facts that are available to us from the question:

1. Pizzeria Roma charges **$14** for a large plain Pizza.
2. Pizzeria Roma charges **$1** for each topping.
3. Pizzeria Tetova charges **$12** for a large plain Pizza.
4. Pizzeria Tetova charges **$1.50** for each topping.
5. Same amount of money for their Pizza.
6. Same number of toppings on their Pizza.

These were the facts or conditions that must be satisfied, and we need to calculate the cost of the Pizza and the number of toppings on the Pizza. What we need to find, in fact, are our unknown variables. Everything else, such as the price for the Pizza and the toppings remain the same, they are constant.

Let's use:

- x -for the number of toppings on the Pizza, and
- y -for the total cost of the Pizza.

Let's first determine how we would calculate the cost of a Pizza from Pizzeria Roma. So, we will order **1** Pizza for **$14** and x toppings priced at **$1** each. In Math notation, the total cost **y** will be:

$$y = 1 \times \$14 + x \times \$1$$

Or simply:

$$y = 14 + x$$

Let's now determine how we would calculate the cost of a Pizza from Pizzeria Tetova. So, you will order **1** Pizza for **$12** and **x** toppings priced at **$1.5** each. In Math notation, the total cost **y** will be:

$$y = 1 \times \$12 + x \times \$1.5$$

Or simply:

$$y = 12 + 1.5x$$

Bear in mind that depending on the number of toppings the total price for the Pizza will be different. In Math this means that **x** and **y** of Pizzeria Roma will be different from the **x** and **y** of Pizzeria Tetova. However, it is possible that for certain number of toppings, even though the price for the Pizza is different, the overall price for Pizza with toppings to be the same. This is the case with Nina's and Artie's Pizzas, and it is given as a condition (see facts **5** and **6**) that they be the same. This means that **x** and **y** of the first equation are the same **x** and **y** of the second equation, meaning they will have the same value and our System of Equations will be:

$$\begin{cases} y = 14 + x \\ y = 12 + 1.5x \end{cases}$$

Let's find **x** and **y** using the Substitution method. We can use the value of **y** of the first equation and substitute it in the second equation. The second equation is:

$$y = 12 + 1.5x$$

After substitution:

$$14 + x = 12 + 1.5x$$

Or we can say that since the left sides of equations (**y**-s) are the same, so must be the right sides of the equations.

Let's solve for **x** :

$$14 + x - 1.5x = 12$$
$$x - 1.5x = 12 - 14$$
$$-0.5x = -2$$
$$x = \frac{-2}{-0.5}$$
$$x = 4$$

Let's substitute the value of **x** in the second equation to find the value of **y**:

$$y = 12 + 1.5x$$
$$y = 12 + 1.5 \times 4$$
$$y = 18$$

So, we found that each Pizza cost **$18** and each Pizza had **4** toppings.

Check your work

Let's use **x = 4** and **y = 18** and confirm that our equations are in balance.

Our System of Equations is:

$$\begin{cases} y = 14 + x \\ y = 12 + 1.5x \end{cases}$$

After we plug in the values for **x** and **y**:

$$\begin{cases} 18 = 14 + 4 \\ 18 = 12 + 1.5 \times 4 \end{cases}$$

$$\begin{cases} 18 = 18 \\ 18 = 18 \end{cases}$$

This proves that our calculations were correct.